# Computer Systems

Ata Elahi

# Computer Systems

## Digital Design, Fundamentals of Computer Architecture and ARM Assembly Language

Second Edition

 Springer

Ata Elahi
Southern Connecticut State University
New Haven, CT, USA

ISBN 978-3-030-93448-4       ISBN 978-3-030-93449-1   (eBook)
https://doi.org/10.1007/978-3-030-93449-1

This Springer imprint is published by the registered company Springer Nature Switzerland AG
The registered company address is: Gewerbestrasse 11, 6330 Cham, Switzerland

*This book is dedicated to Sara, Shabnam, and Aria.*

# Preface

This textbook is the result of my experiences teaching computer systems at the Computer Science Department at Southern Connecticut State University since 1986. The book is divided into three sections: Digital Design, Introduction to Computer Architecture and Memory, and ARM Architecture and Assembly Language. The Digital Design section includes a laboratory manual with 15 experiments using Logisim software to enforce important concepts. The ARM Architecture and Assembly Language section includes several examples of assembly language programs using Keil μVision 5 development tools.

## Intended Audience

This book is written primarily for a one-semester course as an introduction to computer hardware and assembly language for students majoring in Computer Science, Information Systems, and Engineering Technology.

### Changes in the Second Edition
The expansion of Chap. 1 by adding history of computer and Types of Computers. Expanded Chap. 6 "Introduction to Computer Architecture" by adding Computer Abstraction Layers and CPU Instruction Execution Steps. The most revision done on ARM Architecture and Assembly Language by incorporating Keil μvision5, reordering Chaps. 9 and 10, and adding Chap. 11 "C Bitwise and Control Structures used for Programming with C and ARM Assembly Language."

## Organization

The material of this book is presented in such a way that no special background is required to understand the topics.

*Chapter 1–Signals and Number Systems*: Analog Signal, Digital Signal, Binary Numbers, Addition and Subtraction of binary numbers, IEEE 754 Floating Point representations, ASCII, Unicode, Serial Transmission, and Parallel Transmission

*Chapter 2–Boolean Logics and Logic Gates*: Boolean Logics, Boolean Algebra Theorems, Logic Gates, Integrated Circuit (IC), Boolean Function, Truth Table of a function and using Boolean Theorems to simplify Boolean Functions

*Chapter 3–Minterms, Maxterms, Karnaugh Map (K-Map) and Universal Gates*: Minterms, Maxterms, Karnaugh Map (K-Map) to simplify Boolean Functions, Don't Care Conditions and Universal Gates

*Chapter 4–Combinational Logic*: Analysis of Combination Logic, Design of Combinational Logic, Decoder, Encoder, Multiplexer, Half Adder, Full Adder, Binary Adder, Binary Subtractor, Designing Arithmetic Logic Unit (ALU), and BCD to Seven Segment Decoder

*Chapter 5–Synchronous Sequential Logic*: Sequential Logic such as S-R Latch, D-Flip Flop, J-K Flip Flop, T-Flip Flop, Register, Shift Register, Analysis of Sequential Logic, State Diagram, State Table, Flip Flop Excitation Table, and Designing Counter

*Chapter 6–Introduction to Computer Architecture*: Components of a Microcomputer, CPU Technology, CPU Architecture, Instruction Execution, Pipelining, PCI, PCI Express, USB, and HDMI

*Chapter 7–Memory*: Memory including RAM, SRAM, DISK, SSD, Memory Hierarchy, Cache Memory, Cache Memory Mapping Methods, Virtual Memory, Page Table, and the memory organization of a computer

*Chapter 8– Assembly Language and ARM Instructions Part I*: ARM Processor Architecture, and ARM Instruction Set such as Data Processing, Shift, Rotate, Unconditional Instructions and Conditional Instructions, Stack Operation, Branch, Multiply Instructions, and several examples of converting HLL to Assembly Language.

*Chapter 9–ARM Assembly Language Programming Using Keil Development Tools:* Covers how to use Keil development software for writing assembly language using ARM Instructions, Compiling Assembly Language, and Debugging

*Chapter 10–ARM Instructions Part II* and Instruction Formats: This chapter is the continuation of Chap. 8 which covers Load and Store Instructions, Pseudo Instructions, ARM Addressing Mode, and Instruction formats.

*Chapter 11–C Bitwise and Control Structures Used for Programming with C and ARM Assembly Language*

*Instruction Resources*: The instruction resources contain

- 15 Laboratory experiments using Logisim.
- Solutions to the problems of each chapter.
- Power points of each chapter

New Haven, CT, USA                                                                                    Ata Elahi

# Acknowledgments

I would like to express my special thanks to Professor Lancor Chairman of Computer Science Department at Southern Connecticut State University for her support as well as Professor Herv Podnar for his guidance.

I wish to acknowledge and thank Ms. Mary E. James, Senior Editor in Applied Sciences and her assistant, Ms. Zoe Kennedy, for their support.

My special thanks to Eric Barbin, Alex Cushman, Marc Gajdosik, Nickolas Santini, Nicholas Bittar, Omar Abid, and Alireza Ghods for their help in developing the manuscript. Finally, I would like to thank the students of CSC 207 Computer Systems of Spring 2020.

# Contents

# Chapter 1
# Signals and Number Systems

**Objectives: After Completing this Chapter, you Should Be Able to**
- Explain the basic components of a computer.
- Learn the historical development of the computer.
- Represent the hardware and software components of a computer.
- List different types of computers.
- Distinguish between analog and digital signal.
- Learn the characteristics of signal.
- Convert decimal numbers to binary and vice versa.
- Learn addition and subtraction of binary numbers.
- Represent floating numbers in binary.
- Convert from binary to hexadecimal and vice versa.
- Distinguish between serial and parallel transmission.

## 1.1 Introduction

Numerical values have become an integral part of our daily lives. Numerical values can be represented by analog or digital; examples include an analog watch, digital watch, or thermometer. The following are advantages of digital representation of numerical values compared to analog representation:

1. Digital representation is more accurate.
2. Digital information are easier to store.
3. Digital systems are easier to design.
4. Noise has less effect.
5. Digital systems can easily be fabricated in an integrated circuit.

A digital signal is a discrete signal (step by step), and an analog signal is a continuous signal. Digital systems are widely used and its applications can be seen in

© The Author(s), under exclusive license to Springer Nature Switzerland AG 2022
A. Elahi, *Computer Systems*, https://doi.org/10.1007/978-3-030-93449-1_1

**Fig. 1.1** Basic components
of a computer

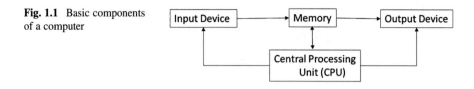

computers, calculators, and cell phones. In a digital system, information is transferred between components of the digital system in the form of digital signals.

A computer consists of two components: hardware and software. Hardware refers to the physical components of a computer such as the keyboard, CPU, and memory. Software refers to programs run by CPU including operating systems and application programs. Computers can come in several different forms such as a desktop, laptop, tablet, server, and iPhone. Regardless of the form, all computers consist of the same basic structures. Figure 1.1 shows the basic components of a computer.

## 1.1.1   CPU

### 1.1.1.1   CPU Execute Program

Input Device

The input device is used for entering information into memory. Examples of input devices include the keyboard, mouse, touch pad screen, light pen, barcode reader, and scanner. The input device converts information into bits, and the bits are stored in memory.

Output Device

A computer's memory transfers information to the *output device* in the form of bits such as the output device converts bits to characters, images, and voices which can be interpreted by humans.

Memory

Memory is used to store information and programs. Memory comes in the form of solid-state electronics such as RAM, ROM, flash drive, or hard disk.

## 1.2   Historical Development of the Computer

**The historical development of the computer can be divided into distinct generations. The first generation of computer was constructed using vacuum tubes and was known as** ENIAC. ENIAC **was developed by** John Mauchly and J. Presper Eckert at the University of Pennsylvania between 1945 and 1953. Figure 1.2 shows an image of a vacuum tube.

**The second generation of computer used transistors and were in use between 1954 and 1965. This generation of computers included the** IBM 7094 (scientific) and the Digital Equipment Corporation (DEC) PDP-1. An image of a transistor is shown in Fig. 1.2.

**The third generation of computer used integrated circuits (IC). These computers were developed between 1965 and 1980, and included the** IBM 360, DEC PDP-8 and PDP-11, and Cray-1 supercomputer. An image of an integrated circuit is shown in Fig. 1.2.

**The fourth generation of computer used VLSI** (Very Large Scale Integration), an evolution of IC technology. This started around 1980 and can be found in processors like the Intel 8080. Figure 1.2 shows an image of VLSI.

## 1.3   Hardware and Software Components of a Computer

The hardware part of a computer is used for the execution of different types of software. Figure 1.3 shows that the hardware is the lowest level component of the computer with different types of software running on top of it.

**Hardware**: The hardware consists of the processor, memory, and I/O controllers.

**System Software**: The system software consists of the compiler and operating system.

Vacuum Tube       Transistor       Integrated Circuit (IC)    VLSI

**Fig. 1.2**  Images of Vacuum Tube, Transistor, IC, and VLSI

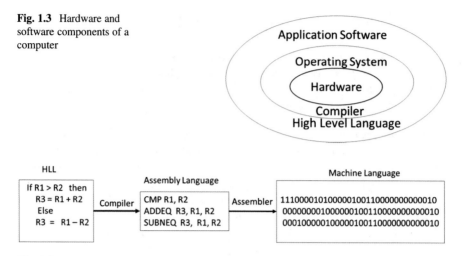

**Fig. 1.3** Hardware and software components of a computer

**Fig. 1.4** Process converting HLL to machine code

**Application Software**: Application software consists of High Level Languages (HLLs) and application software such as Microsoft Office.

**Compiler**: The compiler converts HLL to assembly language, and then the assembler converts that assembly language to machine code (binary) as shown in Fig. 1.4.

**Operating System**: An operating system runs on top of a computer's hardware. Operating systems manage computer hardware resources such as input/output operations, managing memory, and scheduling processes for execution. Some of the most popular operating systems are Windows, MacOS, and Linux.

## 1.4   Types of Computers

As mentioned before, computers come in different forms which can target specific applications. They are:

**Personal Computer (PC):** Personal computers are used by individuals and come with a keyboard and display.

**Server**: A server is a computer with a more powerful CPU than a PC, having larger memory that supports the execution of large programs. Multiple users can access this kind of computer.

**Embedded Computer:** An embedded computer is a computer located inside a device which is used for controlling the operation of the device through a fixed program. Embedded computers can be found in many devices such as dishwashers, laundry machines, automobiles, and robots.

**Supercomputer**: A supercomputer is a computer with many CPUs for running big programs such as weather prediction.

**Cloud computer**: Cloud computing is the delivery of on-demand computing services to clients. A cloud consists of a number of servers, and many clients can access the cloud through an Internet connection for receiving service.

**Personnel Mobile Device (PMD)**: Personal mobile devices are products like smartphones and tablets which can access a server and download information or be used for web browsing with a wireless connection.

## 1.5   Analog Signals

An analog signal is a signal whose amplitude is a function of time and changes gradually as time changes. Analog signals can be classified as nonperiodic and periodic signals.

**Nonperiodic Signal**

In a nonperiodic signal, there is no repeated pattern in the signal as shown in Fig. 1.5.

**Periodic Signal**

A signal that repeats a pattern within a measurable time period is called a periodic signal, and completion of a full pattern is called a *cycle*. The simplest periodic signal is a sine wave, which is shown in Fig. 1.6. In the time domain, the sine wave amplitude $a(t)$ can be represented mathematically as $a(t) = A \, Sin(\omega t + \theta)$ where A is the maximum amplitude, $\omega$ is the angular frequency, and $\theta$ is the phase angle.

An electrical signal, usually representing voice, temperature, or a musical sound, is made of multiple wave forms. These signals have one fundamental frequency and multiple frequencies that are called harmonics.

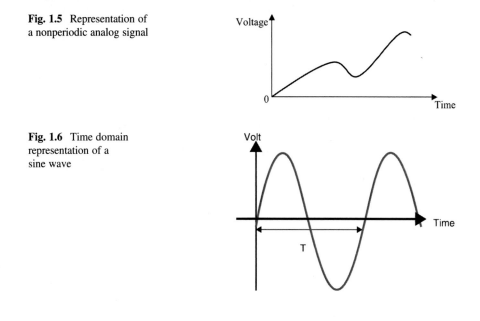

**Fig. 1.5**  Representation of a nonperiodic analog signal

**Fig. 1.6**  Time domain representation of a sine wave

### 1.5.1  Characteristics of an Analog Signal

The characteristics of a periodic analog signal are frequency, amplitude, and phase.
**Frequency**
Frequency (F) is the number of cycles in 1 s, $F = \frac{1}{T}$, where T is time of one cycle in second and $F$ is frequency $i$ represented in Hz (Hertz). If each cycle of an analog signal is repeated every 1 s, the frequency of the signal is 1 Hz. If each cycle of an analog signal is repeated 1000 times every second (once every millisecond), the frequency is

$$f = \frac{1}{T} = \frac{1}{10^{-3}} = 1000 \ \text{Hz} = 1 \ \text{kHz}$$

Table 1.1 shows different values for frequency and their corresponding periods.
**Amplitude**
The amplitude of an analog signal is a function of time as shown in Fig. 1.7 and may be represented in volts (unit of voltage). In other words, the amplitude is its voltage value at any given time. At the time $t_1$, the amplitude of the signal is $V_1$.

**Table 1.1**  Typical units of frequency and period

| Units of frequency | Numerical value | Units of period | Numerical value |
|---|---|---|---|
| Hertz (Hz) | 1 Hz | Second (s) | 1 s |
| Kilohertz (kHz) | $10^3$ Hz | Millisecond (ms) | $10^{-3}$ s |
| Megahertz (MHz) | $10^6$ Hz | Microsecond (μs) | $10^{-6}$ s |
| Gigahertz (GHz) | $10^9$ Hz | Nanosecond (ns) | $10^{-9}$ s |
| Terahertz (THz) | $10^{12}$ Hz | Picosecond (ps) | $10^{-12}$ s |

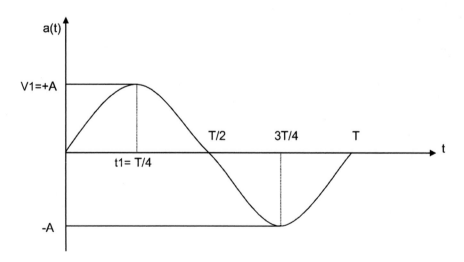

**Fig. 1.7**  A sine wave signal over one cycle

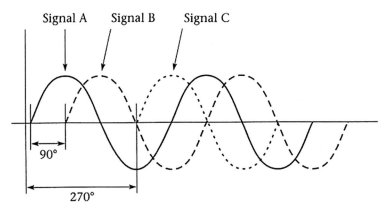

**Fig. 1.8** Three sine waves with different phases

**Phase**

Two signals with the same frequency can differ in phase. This means that one of the signals starts at a different time from the other one. This difference can be represented in degrees (0° to 360°) or by radians. A phase angle of 0° indicates that the sine wave starts at time 0, and a phase angle of 90° indicates that the signal starts at 90° as shown in Fig. 1.8.

**Example 1.1** Find the equation for a sine wave signal with a frequency of 10 Hz, maximum amplitude of 20 V, and phase angle of 0°:

$$\omega = 2\pi f = 2 \times 3.1416 \times 10 = 62.83 \frac{\text{rad}}{\text{s}}$$
$$a(t) = 20 \sin(62.83 \ t)$$

## 1.6   Digital Signals

Modern computers communicate by using digital signals. *Digital signals* are represented by two voltages: one voltage represents the number 0 in binary, and the other voltage represents the number 1 in binary. An example of a digital signal is shown in Fig. 1.9, where 0 volts represents 0 in binary and +5 volts represents 1. 0 or 1 is called a bit and 8 bits is called a byte.

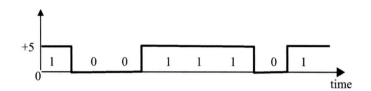

**Fig. 1.9** Digital signal

## 1.7   Number System

Numbers can be represented in different bases. A base of ten is called a decimal. In the example, below consider 356 in decimal:

$$356 = 6 + 50 + 300 = 6 * 10^0 + 5 * 10^1 + 3 * 10^2$$

In general, a number can be represented in the form:

$$(a_5a_4a_3a_2a_1a_0.a_{-1}a_{-2}a_{-3})_r,$$

where $r$ is the base of the number and $a_i$ must be less than $r$.
  $(10011)_2$ is a valid number but $(211.01)_2$ is not.
  Equation 1.1 can be used to convert a number in a given base to decimal:

$$\left( \underbrace{a_5a_4a_3a_2a_1a_0}_{\text{Integer}}.\underbrace{a_{-1}a_{-2}a_{-3}}_{\text{Fraction}} \right)_r = a_0 \times r^0 + a_1 \times r^1 + a_2 \times r^2 + a_3 \times r^3 + \ldots$$
$$+ a_{-1} \times r^{-1} + a_{-2} \times r^{-2} + \cdot a_{-2} \times r^{-3} \ldots$$

$$(1.1)$$

**Example 1.2**  Convert $(27.35)_8$ to the base of 10.

$$(27.35)_8 = 7 * 8^0 + 2 * 8^1 + 3 * 8^{-1} + 5 * 8^{-2} = 7 + 16 + 0.375 + 0.078125$$
$$= (23.45)_{100}$$

**Example 1.3**  Convert 1101111 to decimal.

$$(1101111)_2 = 1 * 2^0 + 1 * 2^1 + 1 * 2^2 + 1 * 2^3 + 0 * 2^4 + 1 * 2^5 + 1 * 2^6$$
$$= 1 + 2 + 4 + 8 + 32 + 64 = (111)_{10}$$

### *1.7.1  Converting from Binary to Decimal*

Equation 1.2 represents the general form of a binary number:

$$(a_5 a_4 a_3 a_2 a_1 a_0 . a_{-1} a_{-2} a_{-3})_2 \tag{1.2}$$

where $a_i$ is a binary digit or bit (either 0 or 1).

Equation 1.2 can be converted to decimal number by using Eq. 1.1:

$$\left( \underbrace{a_5 a_4 a_3 a_2 a_1 a_0}_{\text{Integer}} . \underbrace{a_{-1} a_{-2} a_{-3}}_{\text{Fraction}} \right)_2 = a_0 \times 2^0 + a_1 \times 2^1 + a_2 \times 2^2 + a_3 \times 2^3 + \dots$$
$$+ a_{-1} \times 2^{-1} + a_{-2} \times 2^{-2} + \dots$$

$$\tag{1.3}$$

$$(a_5 a_4 a_3 a_2 a_1 a_0 . a_{-1} a_{-2} a_{-3})_2 = a_0 + 2\ a_1 + 4\ a_2 + 8\ a_3 + 16\ a_4 + 32\ a_5 + 64\ a_6$$
$$+ \frac{1}{2} * a_{-1} + \frac{1}{4} * a_{-2} + \frac{1}{8} * a_{-3}$$

**Example 1.4**  Convert $(110111.101)_2$ to decimal.

$(110111.101)_2 = 1 * 2^0 + 1 * 2^1 + 1 * 2^2 + 0 * 2^3 + 1 * 2^4 + 1 * 2^{*5} + 1 * 2^{-1} + 0 * 2^{-2} + 1 *$
$= 55.625$

*Or*

| $2^5$ | $2^4$ | $2^3$ | $2^2$ | $2^1$ | $2^0$ | | $2^{-1}$ | $2^{-2}$ | $2^{-3}$ |
|---|---|---|---|---|---|---|---|---|---|
| 1 | 1 | 0 | 1 | 1 | 1 | . | 1 | 0 | 1 |

$$32 + 16 + 0 + 4 + 2 + 1 + 1/2 + 0 + 1/8$$

If a binary value is made of $n$ bits of ones, then its decimal value is $2^n - 1$.

**Example 1.5**

$$11 = 2^2 - 1 = 3$$
$$111 = 2^3 - 1 = 7$$
$$1111 = 2^4 - 1 = 15$$
$$11111 = 2^5 - 1 = 31$$
$$111111 = 2^6 - 1 = 63$$

*Binary*, or base of 2 numbers, is represented by 0 s and 1 s. A binary digit, 0 or 1, is called a bit, 8 bits is called a byte, 16 bits is called a half word, and 4 bytes is called a word.

### 1.7.2   Converting from Decimal Integer to Binary

To convert an integer number from decimal to binary, divide the decimal number by the new base (2 for binary), which will result in a quotient and a remainder (either 0 or 1). The first remainder will be the least significant bit of the binary number. Continually divide the quotient by the new base, while taking the remainders as each subsequent bit in the binary number, until the quotient becomes 0.

**Example 1.6**  Convert 34 in decimal to binary.

|        | Quotient | Remainder |
|--------|----------|-----------|
| 34/2 = | 17       | $0 = a_0$ |
| 17/2=  | 8        | $1 = a_1$ |
| 8/2    | 4        | $0 = a_2$ |
| 4/2    | 2        | $0 = a3$  |
| 2/2    | 1        | $0 = a_4$ |
| 1/2    | 0        | $1 = a_5$ |

Therefore $34 = (100010)_2$

If a binary number is made of all ones, then by using the equation $2^n - 1$, it can be converted to decimal.

Examples

| Binary number | $2^n - 1$ | Decimal number |
|---------------|-----------|----------------|
| 11            | $2^2 - 1$ | 3              |
| 111           | $2^3 - 1$ | 7              |
| 1111          | $2^4 - 1$ | 15             |
| 11111         | $2^5 - 1$ | 32             |

A binary number is represented by $a_5\ a_4\ a_3\ a_2\ a_1\ a_0$ where $a_0$ is $2^0$, $a_1$ is $2^1$, and $a_5$ is $2^5$. Table 1.2 shows $2^n$.

### 1.7.3   Converting Decimal Fraction to Binary

A decimal number representation of $(0.XY)_{10}$ can be converted into base of 2 and represented by $(0.a_{-1},\ a_{-2},\ a_{-3},\ \text{etc.})_2$.

**Table 1.2** $2^n$ with different values of n

| $2^n$ | Decimal value | $2^n$ | Decimal value | $2^n$ | Decimal value |
|---|---|---|---|---|---|
| $2^0$ | 1 | $2^8$ | 256 | $2^{16}$ | $65,536 = 64$ K |
| $2^1$ | 2 | $2^9$ | 512 | $2^{17}$ | $131,072 = 128$ K |
| $2^2$ | 4 | $2^{10}$ | $1024 = 1$ K | $2^{18}$ | $262,144 = 256$ K |
| $2^3$ | 8 | $2^{11}$ | $2048 = 2$ K | $2^{19}$ | $524,288 = 512$ K |
| $2^4$ | 16 | $2^{12}$ | $4096 = 4$ K | $2^{20}$ | $1,048,576 = 1$ M |
| $2^5$ | 32 | $2^{13}$ | $8192 = 8$ K | $2^{21}$ | 2 M |
| $2^6$ | 64 | $2^{14}$ | $16,384 = 16$ K | $2^{22}$ | 4 M |
| $2^7$ | 128 | $2^{15}$ | $32,768 = 32$ K | $2^{23}$ | 8 M |

The fraction number is multiplied by 2, the result of integer part is $a_{-1}$ and fraction part multiply by 2, and then separate integer part from fraction, the integer part represents $a_{-2}$; this process continues until the fraction becomes 0.

$(0.35)_{10} = ($            $)_2$

$$0.35 * 2 = 0.7 = 0 + 0.7 \quad a_{-1} = 0$$
$$0.7 * 2 = 1.4 = 1 + 0.4 \quad a_{-2} = 1$$
$$0.4 * 2 = 0.8 = 0 + 0.8 \quad a_{-3} = 0$$
$$0.8 * 2 = 1.6 = 1 + 0.6 \quad a_{-4} = 1$$
$$0.6 * 2 = 1.2 = 1 + 0.2 \quad a_{-5} = 1$$

Sometimes, the fraction does not reach 0 and the number of bits use for the fraction depends on the accuracy that the user defines, therefore $0.35 = 0.010011$ in binary.

The hexadecimal number system has a base of 16 and therefore has 16 symbols (0 through 9, and A through F). Table 1.3 shows the decimal numbers, their binary values from 0 to 15, and their hexadecimal equivalents.

### 1.7.4 Converting from Hex to Binary

Table 1.3 can also be used to convert a number from hexadecimal to binary and from binary to hexadecimal.

**Example 1.7** Convert the binary number 001010011010 to hexadecimal. Each 4 bits are grouped from right to left. By using Table 2.2, each 4-bit group can be converted to its hexadecimal equivalent.

$$\begin{array}{ccc} 0010 & 1001 & 1010 \\ \mathbf{2} & \mathbf{9} & \mathbf{A} \end{array}$$

**Table 1.3** Decimal numbers with binary and hexadecimal equivalents

| Decimal | Binary (base of 2) | Hexadecimal (base of 16) or HEX |
| --- | --- | --- |
| 0 | 0000 | 0 |
| 1 | 0001 | 1 |
| 2 | 0010 | 2 |
| 3 | 0011 | 3 |
| 4 | 0100 | 4 |
| 5 | 0101 | 5 |
| 6 | 0110 | 6 |
| 7 | 0111 | 7 |
| 8 | 1000 | 8 |
| 9 | 1001 | 9 |
| 10 | 1010 | A |
| 11 | 1011 | B |
| 12 | 1100 | C |
| 13 | 1101 | D |
| 14 | 1110 | E |
| 15 | 1111 | F |

**Example 1.8** Convert $(3D5)_{16}$ to binary. By using Table 2.2, the result in binary is

$$3 \quad\quad D \quad\quad 5$$
$$\mathbf{0011 \quad 1101 \quad 0101}$$

The resulting binary number is 001111010101.

**Example 1.9** Convert 6DB from hexadecimal to binary. By using Table 1.3, the result in binary is

$$6 \quad\quad D \quad\quad B$$
$$0110 \quad 1101 \quad 1011$$

The resulting binary number is 011011011011.

**Example 1.10** Convert $(110111.101)_2$ to decimal.

$$(110111.101)_2 = 1*2^0 + 1*2^1 + 1*2^2 + 0*2^3 + 1*2^4 + 1*2^{*5} + 1*2^{-1}$$
$$+ 0*2^{-2} + 1*2^{-3}$$
$$= 55.625$$

### *1.7.5  Binary Addition*

$$1 + 0 = 1, 1 + 1 = 10,$$

Carry bits
$$\begin{array}{r} 111 \\ \mathbf{10101} \\ \mathbf{+01101} \\ \hline \mathbf{100010} \end{array}$$

In a binary number, the first bit from the left of number is called the *most significant bit (MSb)*, and the first bit from the right of number is called *least significant bit (LSb)*.

$$\underline{\text{MSb}} \rightarrow 10010 \leftarrow \underline{\text{LSb}}$$

## 1.8  Complement and Two's Complement

**The idea of the complement makes it simple for a digital computer to perform subtraction and addition.** The complement of 1 is 0 and the complement of 0 is 1.

The complement of a binary number is calculated by complementing each bit of the number.

**Example 1.11**  The complement of 101101 is 010010.

**Two's Complement of a number = Complement of a number + 1**

**Example 1.12**  The two's complement of 101011 is

$$010100(\text{complement}) + 1 = 010101$$

**Example 1.13**  Find the two's complement of 10000:

$$01111(\text{complement}) + 1 = 10000$$

### 1.8.1   Subtraction of Unsigned Number Using Two's Complement

The following procedure is used to subtract $B = b_5\, b_4\, b_3\, b_2\, b_1\, b_0$ from $A = a_5\, a_4\, a_3\, a_2\, a_1\, a_0$:

1. Add two's complement of B to the A.
2. Check to see if the result produces a carry:

   (a) If the result produces a carry, discard the carry and the result is positive.
   (b) If the result does not produce a carry, take two's complement of the result, and the result is negative.

**Example 1.14**   Subtract $B = 101010$ from $A = 110101$.

$$010101 = \text{Complement of B}$$

Two's complement of $B = 010101 + 1 = 010110$.
Add two's complement of B to A.

```
      110101
  +   010110
  ----------------
      10 01011
```

Carry, discard the carry and the result is +001011.

**Example 1.15**   Subtract $B = 110101$ from $A = 101010$.
Two's complement of B is $001010 + 1 = 001011$.
Add two's complement of B to A.

$$
\begin{array}{r}
001011 \\
+101010 \\
\hline
110101
\end{array}
$$

As we can see, adding two 6-bit number results in a 6-bit answer. There is no carry; we just take the two's complement of the result.

$$\text{Two's Complement of } 110101 = 001010 + 1 = -001011$$

## 1.9   Unsigned, Signed Magnitude, and Signed Two's Complement Binary Number

A binary number can be represented in form unsigned number or signed number or signed two's complement, + sign represented by 0 and − sign represented by 1.

### 1.9.1   Unsigned Number

In an unsigned number, all bits of a number are used to represent the number, but in a signed number, the most significant bit of the number represents the sign. A 1 in the most significant position of number represents a negative sign, and 0 in the most significant position of number represents a positive sign.

The 1101 unsigned value is 13.

### 1.9.2   Signed Magnitude Number

In a signed number, the most significant bit represents the sign, where $1101 = -5$ or $0101 = +5$.

In unsigned number, $1101 = 13$.

### 1.9.3   Signed Two's Complement

A signed two's complement applies to a negative number. If the sign of the number is one, then the number is represented by signed two's complement.

**Example 1.16**   Representing $(-5)_{10}$ with 4 bits in signed two's complement.

$(-5)_{10}$ in signed number is 1101, then the two's complement of 101 is 011, and by adding sign bit results in 1011 which represents $-5$ in signed two's complement.

**Example 1.17**   Represent $(-23)_{10}$ with an 8-bit signed two's complement.

$(23)_{10} = (1\ 0\ 1\ 1\ 1)_2$ in unsigned base-2

$(\mathbf{1}\ 00\ 10111)_2$ – Extended to signed 8 bits (notice the *MSb* is **1**)

the two's complement of $(0010111)_2$ is $(1101001)_2$

$(11101001)_2$ is $(-23)_{10}$ in *signed* **two's complement.**

## 1.10  Binary Addition Using Signed Two's Complement

The following examples show the results of the addition of two signed numbers:

(a) (+3) + (+4)

Represent both numbers in binary, and the most significant bit represents the sign, and the result is positive.

$$0011 + 0100 = 0111 = +7$$

(b) $(+ 3)_{10} + (-4)_{10}$

$(-4)_{10}$ in signed two's complement is 1100, then

$0011 + 1100 = 1111$; result does not generate carry of then take two's complement of result which is $-1$

$$(-3) + (+5) = 1101 + 0101 = 10010 (\text{discard carry and result is} + 2)$$

(c) $(-7)_{10} + (-5)_{10}$

By representing both numbers in 4-bit signed two's complement,

$$(-7)_{10} = 1001$$
$$(-5)_{10} = 1011$$

$$1001 + 1011 = 10100$$
$$\uparrow$$

Sign  bit

The addition of two negative numbers results positive, and it is called overflow:

$$(+7) + (+6) = (0111) + (0110) = 1101$$

The addition of two positive numbers results negative, and it is called overflow:

**Example 1.18**  A. The following addition using 8-bit signed two's complement $(-38)_{10} + (44)_{10}$

$(-38)_{10}$ in signed two' complement $=11011010$

$(+44)_{10}$ $\qquad\qquad\qquad\qquad +\underline{00101100}$

$\qquad\qquad\qquad\qquad\qquad\qquad 100000110$ discard the carry and  result is

$\qquad\qquad\qquad\qquad\qquad\qquad\quad 00000110 = 6$

B. Add $-38$ to $-44$ using 8 bit signed two's complement
$-38 = 11011010$
   $-44 = 11010100$
-----------
$10101110 = -82$
C. Add $+100$ to $+44$
$100 = 01100100$
$44 = 00101100$
-------------
$10010000$ the sign of result is negative then results produce overflow.

**Addition Overflow**

The following cases result overflow for adding two signed numbers if:

(a) Both numbers are negative, and results of addition become positive:

$$(-A) + (-B) = +C$$

(b) Both numbers are positive, and results of addition become negative:

$$(+A) + (+B) = -C$$

## 1.11 Floating Point Representation

The central processing unit (CPU) typically consists of an arithmetic logic unit (ALU), floating point unit (FLU/FPU), registers, control unit, and the cache memory.

The *arithmetic logic unit* performs integer arithmetic operations such as addition, subtraction, and logic operations such as AND, OR, XOR, etc. *Integers* are whole numbers without fractional components. 1, 2, and 3 are integers, while 0.1, 2.2, and 3.0001 all have fractional components are called floating point numbers.

The *floating point unit* performs floating point operations. Floating point numbers have a sign, a mantissa, and an exponent. The Institute of Electrical and Electronics Engineers (IEEE) developed a standard to represent floating point numbers, referred to as IEEE 754. This standard defines a format for both single (32-bit) and double (64-bit) precision floating point numbers. Decimal floating points are represented by $M \times 10^E$, where M is the signed mantissa and E is the exponent.

## 1.11.1  Single and Double Precision Representations of Floating Point

Floating point numbers in single precision represented by 32 bits are as shown in Fig. 1.10

### 1.11.1.1  Biased Exponent

The biased exponent is the exponent + 127 $(01111111)_2$; therefore, the exponent is represented by a positive number.

### 1.11.1.2  Normalized Mantissa

The mantissa is represented by 1. M, where M is called normalized mantissa; if M = 00101, then mantissa is 1.00101.

**Example 1.19**  Find normalized mantissa and biased exponent of $(111.0000111)_2$.
111.0000111 can be written in the form of $1.110000111 * 2^{10}$
where
M = 110000111
Biased exponent = 10 + 01111111 = 10000001
The representation of 111.0000111 in single precision is

            1bit    8 bits          23 bits
            0       10000001    11000011100000000000000

**Example 1.20**  Convert the following single precision floating point to decimal number.
101111101 11001000000000000000000
S = 1 means mantissa is negative.
Biased exponent = 01111101.
Exponent = 01111101–01111111 = −00000010.
Normalized mantissa = 11001000000000000000000.
Mantissa = 1. 11001000000000000000000.
Decimal number = $1.11001000000000000000000 *2^{-10}$ = 0.01110011.

            1bit         8 bits                      23 bits
            | S | Biased Exponent | Normalized  Mantissa |

**Fig. 1.10**  IEEE 745 floating point single precision (S = represent sign of mantissa. S = 0 means mantissa is positive, and S = 1 means mantissa is negative)

### 1.11.1.3  Double Precision

In order to increase the accuracy of a floating point number, IEEE 745 offers double precision represented by 64 bits as shown in Fig. 1.11.
  Biased exponent = exponent + 1023

**Example 1.21**  Represent 5.75 in IEEE 745 single precision.
  $-15.625 = (1111.101)_2$
  $-1111.101 = -1.11101101 * 2^{11}$
  $S = 1$
  Normalized mantissa $= 0.11101101$.
  Biased exponent $= 11 + 01111111 = 10000010$.
  IEEE745 single precision is 1 10000010 11101101000000000000000.

## 1.12  Binary-Coded Decimal (BCD)

In daily life, we use decimal numbers where the largest digit is 9, which is represented by 1001 in binary. Table 1.4 shows decimal numbers and their corresponding BCD code.

**Example 1.22**  Converting 345 to BCD
  Using the table: 0011 0100 0101

**Example 1.23**  Convert $(10100010010)_{BCD}$ to decimal, separate each 4 bits from right to left, and substitute the corresponding decimal number with BCD the results in 512.

**Fig. 1.11**  IEEE 745 double precision floating point format

| 1bit | 11 bits | 52 bits |
|---|---|---|
| S | Biased Exponent | Normalized Mantissa |

**Table 1.4**  Binary-coded decimal (BCD)

| Decimal | BCD |
|---|---|
| 0 | 0000 |
| 1 | 0001 |
| 2 | 0010 |
| 3 | 0011 |
| 4 | 0100 |
| 5 | 0101 |
| 6 | 0110 |
| 7 | 0111 |
| 8 | 1000 |
| 9 | 1001 |

## 1.13    Coding Schemes

### 1.13.1  ASCII Code

Each character in ASCII code has a representation using 8 bits, where the most significant bit is used for a parity bit. Table 1.5 shows the *ASCII code* and its hexadecimal equivalent.

Characters from hexadecimal 00 to 1F and 7F are control characters which are nonprintable characters, such as NUL, SOH, STX, ETX, ESC, and DLE (data link escape).

**Example 1.24** Convert the word "network" to binary and show the result in hexadecimal. By using Table 1.4, each character is represented by 7 bits and results in:

| 1001110 | 1100101 | 1110100 | 1110111 | 1101111 | 1110010 | 1101011 |
|---------|---------|---------|---------|---------|---------|---------|
| N       | e       | t       | w       | o       | r       | k       |

Or in hexadecimal

| 4E | 65 | 74 | 77 | 6F | 72 | 6B |
|----|----|----|----|----|----|----|

### 1.13.2  Universal Code or Unicode

Unicode is a new 16-bit character-encoding standard for representing characters and numbers in most languages such as Greek, Arabic, Chinese, and Japanese. The ASCII code uses 8 bits to represent each character in Latin, and it can represent 256 characters. The ASCII code does not support mathematical symbols and scientific symbols. Since *Unicode* uses 16 bits, it can represent 65,536 characters or symbols. A character in Unicode is represented by 16-bit binary, equivalent to 4 digits in hexadecimal. For example, the character B in Unicode is U0042H (U represents Unicode). The ASCII code is represented between $(00)_{16}$ and $(FF)_{16}$. For converting ASCII code to Unicode, two zeros are added to the left side of ASCII code; therefore, the Unicode to represent ASCII characters is between $(0000)_{16}$ and $(00FF)_{16}$. Table 1.6 shows some of the Unicode for Latin and Greek characters. Unicode is divided into blocks of code, with each block assigned to a specific language. Table 1.7 shows each block of Unicode for some different languages (Fig. 1.12).

Example of Unicode: open Microsoft Word and click on insert then symbol will result Fig. 1.12. Click on any character to display the Unicode value of the character, for example, Unicode for β is 03B2 in hex.

**Table 1.5**  American Standard Code for Information Interchange (ASCII)

| Binary | Hex | Char | Binary | Hex | Char | Binary | Hex | Char | Binary | Hex | Char |
|---|---|---|---|---|---|---|---|---|---|---|---|
| 0000000 | 00 | NUL | 0100000 | 20 | SP | 1000000 | 40 | @ | 1100000 | 60 | ` |
| 0000001 | 01 | SOH | 0100001 | 21 | ! | 1000001 | 41 | A | 1100001 | 61 | a |
| 0000010 | 02 | STX | 0100010 | 22 | " | 1000010 | 42 | B | 1100010 | 62 | b |
| 0000011 | 03 | ETX | 0100011 | 23 | # | 1000011 | 43 | C | 1100011 | 63 | c |
| 0000100 | 04 | EOT | 0100100 | 24 | $ | 1000100 | 44 | D | 1100100 | 64 | d |
| 0000101 | 05 | ENQ | 0100101 | 25 | % | 1000101 | 45 | E | 1100101 | 65 | e |
| 0000110 | 06 | ACK | 0100110 | 26 | & | 1000110 | 46 | F | 1100110 | 66 | f |
| 0000111 | 07 | BEL | 0100111 | 27 | ' | 1000111 | 47 | G | 1100111 | 67 | g |
| 0001000 | 08 | BS | 0101000 | 28 | ( | 1001000 | 8 | H | 1101000 | 68 | h |
| 0001001 | 09 | HT | 0101001 | 29 | ) | 1001001 | 49 | I | 1101001 | 69 | i |
| 0001010 | 0A | LF | 0101010 | 2A | * | 1001010 | 4A | J | 1101010 | 6A | j |
| 0001011 | 0B | VT | 0101011 | 2B | + | 1001011 | 4B | K | 1101011 | 6B | k |
| 0001100 | 0C | FF | 0101100 | 2C | , | 1001100 | 4C | L | 1101100 | 6C | l |
| 0001101 | 0D | CR | 0101101 | 2D | - | 1001101 | 4D | M | 1101101 | 6D | m |
| 0001110 | 0E | SO | 0101110 | 2E | . | 1001110 | 4E | N | 1101110 | 6E | n |
| 0001111 | 0F | SI | 0101111 | 2F | / | 1001111 | 4F | O | 1101111 | 6F | o |
| 0010000 | 10 | DLE | 0110000 | 30 | 0 | 1010000 | 50 | P | 1110000 | 70 | p |
| 0010001 | 11 | DC1 | 0110001 | 31 | 1 | 1010001 | 51 | Q | 1110001 | 71 | q |
| 0010010 | 12 | DC2 | 0110010 | 32 | 2 | 1010010 | 52 | R | 1110010 | 72 | r |
| 0010011 | 13 | DC3 | 0110011 | 33 | 3 | 1010011 | 53 | S | 1110011 | 73 | s |
| 0010100 | 14 | DC4 | 0110100 | 34 | 4 | 1010100 | 54 | T | 1110100 | 74 | t |
| 0010101 | 15 | NACK | 0110101 | 35 | 5 | 1010101 | 55 | U | 1110101 | 75 | u |
| 0010110 | 16 | SYN | 0110110 | 36 | 6 | 1010110 | 56 | V | 1110110 | 76 | v |
| 0010111 | 17 | ETB | 0110111 | 37 | 7 | 1010111 | 57 | W | 1110111 | 77 | w |
| 0011000 | 18 | CAN | 0111000 | 38 | 8 | 1011000 | 58 | X | 1111000 | 78 | x |

(continued)

**Table 1.5** (continued)

| Binary | Hex | Char | Binary | Hex | Char | Binary | Hex | Char | Binary | Hex | Char |
|---|---|---|---|---|---|---|---|---|---|---|---|
| 0011001 | 19 | EM | 0111001 | 39 | 9 | 1011001 | 59 | Y | 1111001 | 79 | y |
| 0011010 | 1A | SUB | 0111010 | 3A | : | 1011010 | 5A | Z | 1111010 | 7A | z |
| 0011011 | 1B | ESC | 0111011 | 3B | ; | 1011011 | 5B | [ | 1111011 | 7B | { |
| 0011100 | 1C | FS | 0111100 | 3C | < | 1011100 | 5C | \ | 1111100 | 7C | | |
| 0011101 | 1D | GS | 0111101 | 3D | = | 1011101 | 5D | ] | 1111101 | 7D | } |
| 0011110 | 1E | RS | 0111110 | 3E | > | 1011110 | 5E | ^ | 1111110 | 7E | ~ |
| 0011111 | 1F | US | 0111111 | 3F | ? | 1011111 | 5F | _ | 1111111 | 7F | DEL |

**Table 1.6** Unicode values for some Latin and Greek characters

| Latin | | Greek | |
|---|---|---|---|
| Character | Code (hex) | Character | Code (hex) |
| A | U0041 | φ | U03C6 |
| B | U0042 | α | U03B1 |
| C | U0043 | γ | U03B3 |
| 0 | U0030 | μ | U03 BC |
| 8 | U0038 | β | U03B2 |

**Table 1.7** Unicode block allocations

| Start code (hex) | End code (hex) | Block name |
|---|---|---|
| U0000 | U007F | Basic Latin |
| U0080 | U00FF | Latin supplement |
| U0370 | U03FF | Greek |
| U0530 | U058F | Armenian |
| U0590 | U05FF | Hebrew |
| U0600 | U06FF | Arabic |
| U01A0 | U10FF | Georgian |

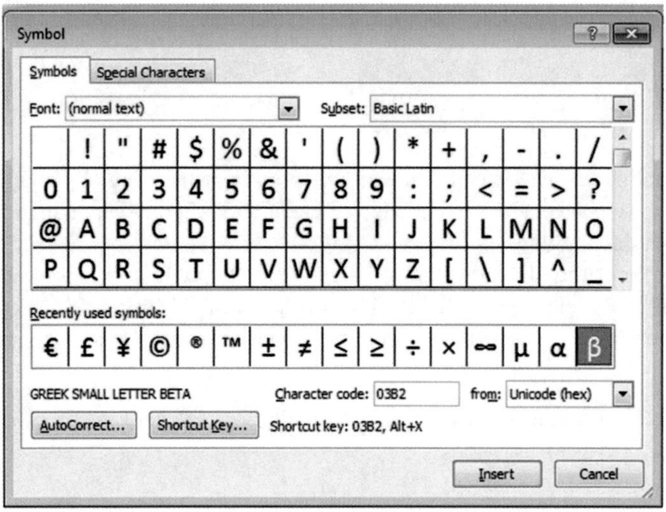

**Fig. 1.12** Example of Unicode

## 1.14   Parity Bit

A parity bit is used for error detection of information, since a bit or bits may be changed during the transmission of information from source to destination, a parity bit is an extra bit appended to the information. It represents whether the number of ones or zeroes is either even or odd in the original transmission and can alert the destination to a loss of information.

### 1.14.1 Even Parity

The extra bit (0 or 1) is chosen such that the number of ones becomes even.

**Example 1.25** Our message is $(00111)_2$. By appending a one to the left side of the message, we create $(100111)_2$. Our *even* parity bit has made the total number of ones even (from 3 to 4 ones).

Our message is $(10111)_2$. By appending a zero to the left side of the message, we create $(010111)_2$. Our *even* parity bit has left the total number of ones even (4 ones).

### 1.14.2 Odd Parity

The extra bit (0 or 1) is chosen such that the number of ones becomes odd.

Our message is $(10111)_2$. By appending a one to the left side of the message, we create $(110111)_2$. Our *odd* parity bit has made the total number of ones even (from 4 to 5 ones).

## 1.15 Clock

0 and 1 continuously repeated is called clock as shown in Fig. 1.13, when clock change from 0 to 1 is called rising edge of clock and when clock change from 1 to 0 is called falling edge of clock.

Each cycle of the clock consists of 1 and 0 or 0 and 1; it is measured by time (second). If one cycle represented by T and the unit of T is seconds, then

F (frequency) = 1/T where the unit of frequency is hertz (Hz) and the unit of T is seconds.

**Example 1.26** What is the frequency of a clock if one cycle of the clock is equal to 0.5 ms?

$$F = 1/T = 1/0.5 \times 10^{-3} = 2000 \ \text{Hz}$$

$$
\begin{array}{ll}
1000 \ \text{Hz} & \text{Kilohertz (KHz)} \\
10^6 \ \text{Hz} & \text{Megahertz (MHz)} \\
10^9 \ \text{Hz} & \text{Gigahertz (GHz)}
\end{array}
$$

**Fig. 1.13** Clock signals

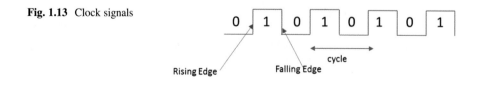

## 1.16  Transmission Modes

When data is transferred from one computer to another by digital signals, the receiving computer has to distinguish the size of each signal to determine when a signal ends and when the next one begins. For example, when a computer sends a signal as shown in Fig. 1.14, the receiving computer has to recognize how many ones and zeros are in the signal. Synchronization methods between source and destination devices are generally grouped into two categories: asynchronous and synchronous.

### *1.16.1  Asynchronous Transmission*

*Asynchronous transmission* occurs character by character and is used for serial communication, such as by a modem or serial printer. In asynchronous transmission, each data character has a start bit which identifies the start of the character and 1 or 2 bits which identifies the end of the character, as shown in Fig. 1.15. The data character is 7 bits. Following the data bits may be a parity bit, which is used by the receiver for error detection. After the parity bit is sent, the signal must return to high for at least 1 bit time to identify the end of the character. The new start bit serves as an indicator to the receiving device that a data character is coming and allows the receiving side to synchronize its clock. Since the receiver and transmitter clock are not synchronized continuously, the transmitter uses the start bit to reset the receiver clock so that it matches the transmitter clock. Also, the receiver is already programmed for the number of bits in each character sent by the transmitter.

**Fig. 1.14**  Digital signals

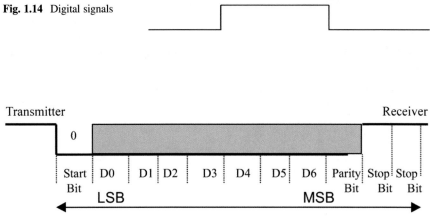

**Fig. 1.15**  Asynchronous transmission

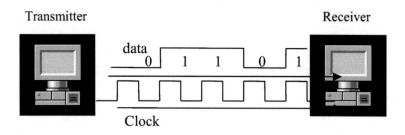

**Fig. 1.16** Synchronous transmission

## *1.16.2 Synchronous Transmission*

Some applications require transferring large blocks of data, such as a file from disk or transferring information from a computer to a printer. *Synchronous transmission* is an efficient method of transferring large blocks of data by using time intervals for synchronization.

One method of synchronizing transmitter and receiver is through the use of an external connection that carries a clock pulse. The clock pulse represents the data rate of the signal, as shown in Fig. 1.16, and is used to determine the speed of data transmission. The receiver of Fig. 1.16 reads the data as 01101, each bit width represented by one clock.

Figure 1.16 shows that an extra connection is required to carry the clock pulse for synchronous transmission. In networking, one medium is used for transmission of both information and the clock pulse. The two signals are encoded in a way that the synchronization signal is embedded into the data. This can be done with Manchester encoding or Differential Manchester encoding.

## 1.17   Transmission Methods

There are two types of transmission methods used for sending digital signals from one station to another across a communication channel: serial transmission and parallel transmission.

## *1.17.1 Serial Transmission*

In *serial transmission*, information is transmitted 1 bit at a time over one wire as shown in Fig. 1.17.

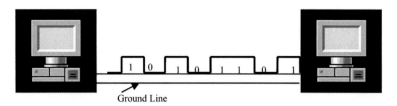

Ground Line

**Fig. 1.17**  Serial transmission

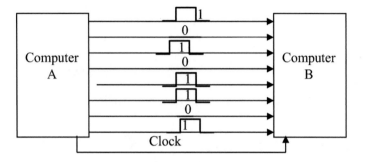

Clock

**Fig. 1.18**  Parallel transmission

## *1.17.2   Parallel Transmission*

In *parallel transmission*, multiple bits are sent simultaneously, 1 byte or more at a time, instead of bit by bit as in serial transmission. Figure 1.18 shows how computer A sends 8 bits of information to computer B at the same time by using eight different wires. Parallel transmission is faster than serial transmission, at the same clock speed.

## 1.18   Summary

- The basic components of a computer are input devices, CPU, output devices, and memory.
- Different forms of computers include the PC, server, embedded computer, super-computer, cloud computer, and PMD.
- The Instruction Set will interface software with hardware.
- Compilers convert HLL to machine code.
- Firmware is a program that can enable a device for an operation such a driver.
- Information exchanges between two electrical systems are in the form of digital or analog signals.
- Modern computers work with digital signals:

- A digital signal is represented by two voltages.
- Binary is the representation of a number in base of 2.
- The frequency of a signal is the inverse of its cycle time.
- One digit in binary is called a bit, 8 bits is called 1 byte, and 4 bytes is called one word.
- Information is represented inside the computer by binary or base of 2.
- Negative numbers inside the computer are represented by the two's complement.
- The most significant bit in signed numbers is represented by the sign of the number.
- Positive is represented by zero and negative is represented by one.
- One's complement of a binary number is the complement of each bit of the number.
- Two's complement of a binary number is the complement of the number plus one.
- Binary-coded decimal (BCD) is used for representing decimal numbers from 0 to 9.
- The hexadecimal number system is used in digital systems and computers in an efficient way of representing binary quantities.
- Parity bit is used for error detection of one bit error.
- IEEE 757 standard is used to represent floating point number.
- Information is represented by ASCII code inside the computer; ASCII code is made of 7 bits.
- Information between components of a computer can be transmitted in serial or parallel form.
- In serial transmission, information is transmitted 1 bit at a time.
- In parallel transmission, information is transmitted in multiple bits at a time.
- The next chapter will cover Boolean logic, Boolean algebra theorems, and logic gates. Logic gate is the basic component of an integrated circuit (IC).

**Problems and Questions**

1. List the components of a computer.
2. List the different types of computers.
3. What is the function of an operating system?
4. List the names of two operating systems.
5. What is the function of compiler?
6. List three computer input devices.
7. List three computer output devices.
8. Show a digital signal.
9. Show an analog signal.
10. How many bits is:

    (a) Byte
    (b) Half word
    (c) Word

11. Convert the following decimal numbers to binary:

   (a) 35
   (b) 85
   (c) 23.25

12. Convert the following binary numbers to decimal

   (a) 111111
   (b) 1010101
   (c) 1101001.101

13. Convert the following binary numbers to decimal:

   (a) 1111101
   (b) 1010111.1011
   (c) 11111111
   (d) 10000000

14. Convert the following binary numbers to hexadecimal:

   (a) 1110011010
   (b) 1000100111
   (c) 101111.101

15. Find the frequency of digital signal with the following clock cycles:

   (a) 1 s
   (b) s
   (c) 0.02 s
   (d) 0.02 ms

16. The following frequencies of a digital signal are given, find the clock cycle of digital signal:

   (a) 10 Hz
   (b) 200 Hz
   (c) 10,000 Hz
   (d) 4 MHz

17. Convert each of the following numbers to base of 10:

   (a) $(34A)_{16}$
   (b) $(FAC)_{16}$

18. Convert the following decimal numbers to base of 16:

   (a) $(234)_{10}$
   (b) $(75)_{10}$

19. Convert the following numbers to binary:

(a) $(3FDA)_{16}$
(b) $(FDA.5F)_{16}$

20. Perform the following additions:

$$1101010 \qquad 1100101$$
$$\underline{1011011} \quad \underline{+1010111}$$

21. Find two's complements of the following numbers:

(a) 11111111
(b) 10110000
(c) 10000000
(d) 00000000

22. The word "LOGIC" is given.

(a) Represent in ASCII.
(b) Add even parity bit to each character and represent each character in hex.

23. Represent $(465)_{10}$ in BCD.
24. Represent $(100101100111)_{BCD}$ in decimal.
25. Convert the following two's complement numbers to decimal:

(a) 1011
(b) 11111001
(c) 10011111

26. Subtract the following unsigned numbers using two's complement:

(a) 11110011–11000011
(b) 10001101–11111000
(c) 11111101–11000001
(d) 10011001–11100001

27. Perform addition of the following signed numbers; assume each number is represented by 6 bits and state if the result of each addition produces overflow:

(a) (+12) + (+7)
(b) (+25) + (+30)
(c) (−5) + (+ 9)
(d) (−6) + (−7)
(e) (−36) + (−12)

28. What is the largest 16-bit binary value that can be represented by:

    (a) Unsigned number
    (b) Signed magnitude
    (c) Signed two's complement

29. Represent the following decimal numbers in IEEE 745 single precision:

    (a) 34.375
    (b) −0.045

30. Convert the following IEEE 745 single precision to decimal number:

    (a) 1 10000100 01110000000000000000000
    (b) 0 01111100 11100000000000000000000

31. List the types of transmission modes.
32. Convert each of the following signed two's complement numbers to decimal:

    (a) 11000011
    (b) 10001111

33. Represent each of the following numbers in 8-bit signed two's complement:

    (a) −15
    (b) −24
    (c) −8

34. Perform the following addition:

    (a) $(0F4A)_{16} + (420B)_{16}$
    (b) $(084C)_{16} + (1265)_{16}$

# Chapter 2
# Boolean Logics and Logic Gates

**Objectives: After Completing this Chapter, you Should Be Able to**
- Understand the basic operation of Boolean theorems.
- Explain the operation of different logic gates such as AND, OR, NOT, XOR, and NAND gates.
- Show the truth table of different logic gates.
- Distinguish between the different types of integrated circuits (ICs).
- Apply Boolean theorems to simplify Boolean function.
- Draw logic circuit for Boolean function.
- Show the truth table of Boolean function.
- Find the output function of a digital logic circuit.
- Distinguish between the SSI, MSI, LSI, and VLSI.

## 2.1   Introduction

Logic gates are made of transistors, and they are the basic components of integrated circuit (IC). Logic gates are used for designing digital system; there are three basic logic operations and they are called AND, OR, and NOT. The characteristic of a digital system can be represented by a function or truth table. Boolean theorems are used to simplify Boolean function in order to use fewer logic gates. Integrated circuits are classified based on the number of gates that they contain, and they are called SSI, MSI, LSI, and VLSI.

A. Elahi, *Computer Systems*, https://doi.org/10.1007/978-3-030-93449-1_2

## 2.2  Boolean Logics and Logic Gates

### 2.2.1  AND Logic

The AND logic is denoted by a period "." but most of the time, the period is left out. X.Y or XY is pronounced as X AND Y.

X AND Y = Z,   Z = 1 if and only if X = 1 and Y = 1 otherwise Z = 0.

The AND logic operation can be represented by the electrical circuit in Fig. 2.1.
Assume X and Y are switches and Z is the light; X = 0, Y = 0 means the switches are open and the light being off means zero and the light on means one. Then we can make a table; Table 2.1 shows the operation of Fig. 2.1.
Figure 2.2 shows a 2-input AND gate and Table 2.2 shows the truth table for the AND gate. The output of the AND gate is one when both inputs are one.

**Fig. 2.1** Representation of the AND operation

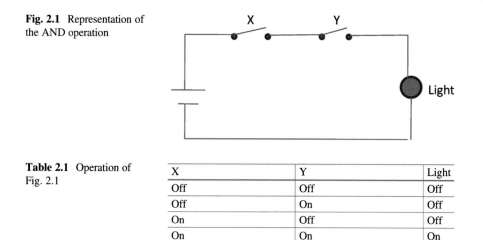

**Table 2.1** Operation of Fig. 2.1

| X | Y | Light |
|---|---|---|
| Off | Off | Off |
| Off | On | Off |
| On | Off | Off |
| On | On | On |

**Fig. 2.2** 2-input AND gate

X
Y   Z=XY

**Table 2.2** AND gate truth table

| X | Y | Z |
|---|---|---|
| 0 | 0 | 0 |
| 0 | 1 | 0 |
| 1 | 0 | 0 |
| 1 | 1 | 1 |

## 2.2.2 OR Logic

The OR operation is represented by a plus sign, "+" or a V, where "+" is the most popular symbol used. X + Y is pronounced X OR Y.

$$X + Y = Z, \quad Z = 1 \text{ if } X = 1 \text{ OR } Y = 1 \text{ or both } X = 1 \text{ and } Y = 1.$$

The OR operation can be represented by the electrical circuit in Fig. 2.3. In Fig. 2.3, the light is off when both switches are off, and the light is on when at least one switch is closed. Figure 2.4 shows a 2-input OR gate and Table 2.3 shows truth table for the 2-input OR gate.

## 2.2.3 NOT Logic

The NOT logic performs a complement, meaning it converts a 1 to 0 and 0 to 1. Also called an inverter, the NOT X is represented by $X'$ or $\overline{X}$. Figure 2.5 shows the NOT gate and Table 2.4 shows truth table for the NOT gate (inverter).

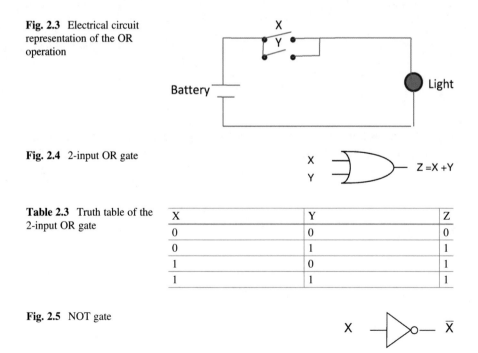

**Fig. 2.3** Electrical circuit representation of the OR operation

**Fig. 2.4** 2-input OR gate

**Table 2.3** Truth table of the 2-input OR gate

| X | Y | Z |
|---|---|---|
| 0 | 0 | 0 |
| 0 | 1 | 1 |
| 1 | 0 | 1 |
| 1 | 1 | 1 |

**Fig. 2.5** NOT gate

**Table 2.4** Truth table for the NOT gate

| X | X' |
|---|-----|
| 0 | 1 |
| 1 | 0 |

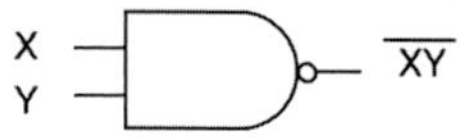

**Fig. 2.6** 2-input NAND gate

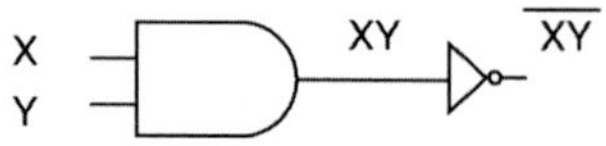

**Fig. 2.7** AND-NOT gates used together to act as NAND

**Table 2.5** Truth table of the 2-input NAND

| X | Y | $\overline{XY}$ |
|---|---|------|
| 0 | 0 | 1 |
| 0 | 1 | 1 |
| 1 | 0 | 1 |
| 1 | 1 | 0 |

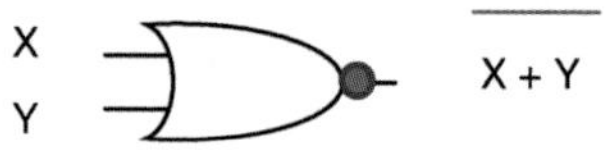

**Fig. 2.8** NOR gate

**Table 2.6** Truth table for the 2-input NOR gate

| X | Y | $\overline{X+Y}$ |
|---|---|--------|
| 0 | 0 | 1 |
| 0 | 1 | 0 |
| 1 | 0 | 0 |
| 1 | 1 | 0 |

## 2.2.4 NAND Gate

Figure 2.6 shows 2-input NAND gate. The NAND gate can be made from an AND gate and a NOT gate as shown in Fig. 2.7. Table 2.5 shows the truth table of the 2-input NAND gate.

## 2.2.5 NOR Gate

Figure 2.8 shows a NOR logic gate. NOR gates are made of OR and NOT gates. Table 2.6 shows the truth table of the 2-input NOR gate.

### 2.2.6   Exclusive OR Gate

Figure 2.9 shows an exclusive OR gate. Exclusive OR is represented by $\oplus$ and is labeled XOR. Table 2.7 shows the truth table for the XOR gate.

$$\mathbf{X} \oplus \mathbf{Y} = \mathbf{X'Y} + \mathbf{XY'}$$

### 2.2.7   Exclusive NOR Gate

Figure 2.10 shows an exclusive NOR gate. Exclusive NOR is represented by $\odot$ and labeled XNOR. Table 2.8 shows the truth table for the exclusive NOR gate.

### 2.2.8   Tri-State Device

Figure 2.11 shows the diagram of a tri-state device; the control line controls the operation of tri-state devices.

**Fig. 2.9**  2-input XOR

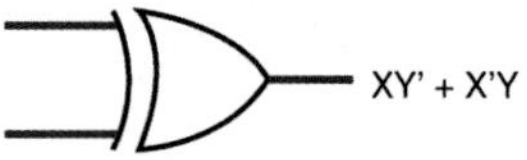

**Table 2.7**  Truth table for XOR gate

| X | Y | X $\oplus$ Y |
|---|---|---|
| 0 | 0 | 0 |
| 0 | 1 | 1 |
| 1 | 0 | 1 |
| 1 | 1 | 0 |

**Fig. 2.10**  Exclusive NOR gate

**Table 2.8**  Truth table for exclusive NOR gate

| X | Y | X $\odot$ Y |
|---|---|---|
| 0 | 0 | 1 |
| 0 | 1 | 0 |
| 1 | 0 | 0 |
| 1 | 1 | 1 |

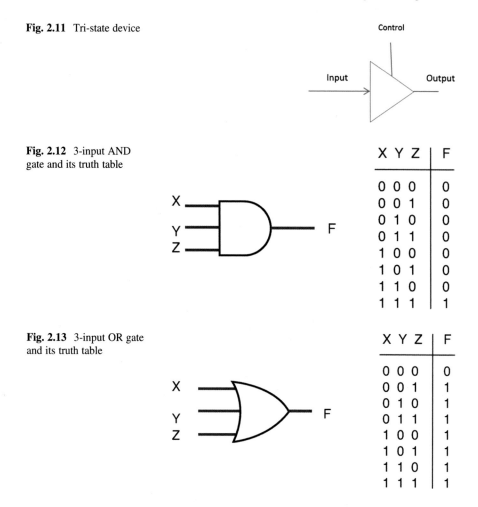

**Fig. 2.11** Tri-state device

**Fig. 2.12** 3-input AND gate and its truth table

**Fig. 2.13** 3-input OR gate and its truth table

| X | Y | Z | F |
|---|---|---|---|
| 0 | 0 | 0 | 0 |
| 0 | 0 | 1 | 0 |
| 0 | 1 | 0 | 0 |
| 0 | 1 | 1 | 0 |
| 1 | 0 | 0 | 0 |
| 1 | 0 | 1 | 0 |
| 1 | 1 | 0 | 0 |
| 1 | 1 | 1 | 1 |

| X | Y | Z | F |
|---|---|---|---|
| 0 | 0 | 0 | 0 |
| 0 | 0 | 1 | 1 |
| 0 | 1 | 0 | 1 |
| 0 | 1 | 1 | 1 |
| 1 | 0 | 0 | 1 |
| 1 | 0 | 1 | 1 |
| 1 | 1 | 0 | 1 |
| 1 | 1 | 1 | 1 |

In Fig. 2.11, if the control line is set to zero, then there is no connection between input and output (the output is high impedance). If the control line is set to one, then output value is equal to the input value.

## 2.2.9  Multiple Inputs Logic Gates

Figure 2.12 shows 3-input AND gate with its truth table, Fig. 2.13 shows 3-input OR gate with its truth table, Fig. 2.14 shows 3-input NOR gates with its truth table, and Fig. 2.15 shows 3-input NAD gate with its corresponding table.

**Fig. 2.14**  3-input NAND
gate and its truth table

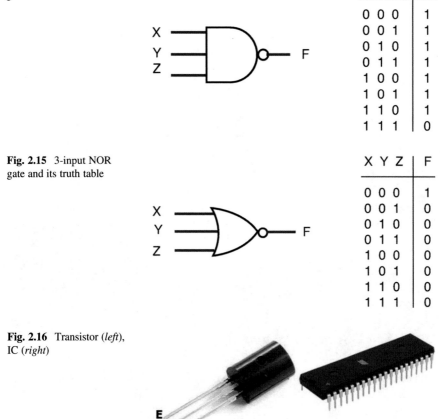

| X | Y | Z | F |
|---|---|---|---|
| 0 | 0 | 0 | 1 |
| 0 | 0 | 1 | 1 |
| 0 | 1 | 0 | 1 |
| 0 | 1 | 1 | 1 |
| 1 | 0 | 0 | 1 |
| 1 | 0 | 1 | 1 |
| 1 | 1 | 0 | 1 |
| 1 | 1 | 1 | 0 |

**Fig. 2.15**  3-input NOR
gate and its truth table

| X | Y | Z | F |
|---|---|---|---|
| 0 | 0 | 0 | 1 |
| 0 | 0 | 1 | 0 |
| 0 | 1 | 0 | 0 |
| 0 | 1 | 1 | 0 |
| 1 | 0 | 0 | 0 |
| 1 | 0 | 1 | 0 |
| 1 | 1 | 0 | 0 |
| 1 | 1 | 1 | 0 |

**Fig. 2.16**  Transistor (*left*),
IC (*right*)

## 2.3    Integrated Circuit (IC) Classifications

A transistor is a basic component of integrated circuits (IC). Figure 2.16 shows a
transistor with an IC. Transistors act like a switch in integrated circuits. An integrated circuit is made from a hundred to millions of transistors.

Integrated circuits are classified based on the number of the gates inside the IC
such as SSI, MSI, LSI, and VLSI.

**Fig. 2.17** TTL 7408 2 input
AND gate

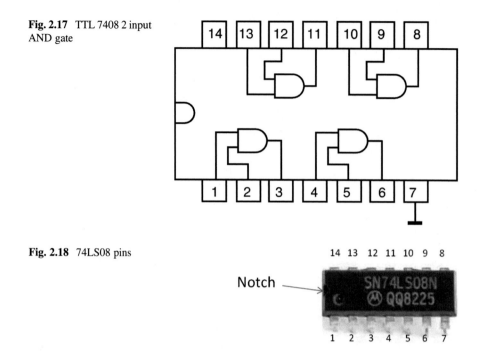

**Fig. 2.18** 74LS08 pins

## 2.3.1  Small-Scale Integration (SSI)

SSI refers to an IC that has less than 10 gates. Figure 2.17 shows the inside of a
74HC08 (7408) IC, and Fig. 2.18 shows an image of a 74HC08 IC.

## 2.3.2  Integrated Circuit Pins Numbering

Figure 2.18 shows a TTL 7408 IC; an IC chip should be read with the notch on the
left-hand side. The labeling convention starts with the left bottom pin under the
notch and goes in a counterclockwise direction.

The left bottom pin is pin #1.
The right bottom pin is pin #7.
The right upper pin is pin #8.
The left upper pin is pin #14.

As shown in Fig. 2.18, the IC number is 74LS08 where LS represents the material
IC was made with. Also on the IC, there is the letter *M* representing Motorola which
is the manufacturer of the IC, Intel uses the character "i," and Texas Instruments uses
a map of Texas.

### 2.3.3   Medium-Scale Integration (MSI)

Refers to an IC that contains between 10 and 100 gates such as decoders and multiplexers.

### 2.3.4   Large-Scale Integration (LSI)

Refers to an IC that contains between 100 and 1000 gates.

### 2.3.5   Very-Large-Scale Integration (VLSI)

Refers to an IC that contains more than 1000 gates.

## 2.4   Boolean Algebra Theorems

Boolean theorems are used to simplify Boolean functions in order to use fewer gates. Any variable such as X in binary can have a value of one or zero.

**Theorem 1**
$$X + X = X$$

*Proof*: Select X as 0 then $0 + 0 = 0$; select $X = 1$ then $1 + 1 = 1$ results: $X + X = X$.

**Theorem 2**
$$X + 1 = 1$$

*Proof*: Select $X = 0$ then $0 + 1 = 1$; select $X = 1$ then $1 + 1 = 1$; both cases result in 1 then $X + 1 = 1$.

**Theorem 3**
$$X + 0 = X$$

*Proof*: Select $x = 0$ then $0 + 0 = 0$; select $X = 1$ then $1 + 0 = 1$; the result is whatever value X is.

**Theorem 4**
$$X + X' = 1$$

*Proof:* Select $X = 0$ then $0 + 1 = 1$; select $X = 1$ then $1 + 0 = 1$; in both cases the result is 1.

**Theorem 5**

$$X.X = X$$

*Proof:* Select $X = 1$ then $1.1 = 1$; select $X = 0$ then $0.0 = 0$; therefore $XX = X$.

**Theorem 6**

$$X.1 = X$$

*Proof:* Select $X = 1$ then $1.1 = 1$; select $X = 0$ then $0.1 = 0$; therefore $X.1 = X$.

**Theorem 7**

$$X.X' = 0$$

*Proof:* Select $X = 0$ then $0.1 = 0$; select $X = 1$ then $1.0 = 0$; both values of $X$ result in 0.

**Theorem 8**

$$(X')' = X$$

$(0')' = (1)' = 0$, $(1')' = (0)' = 1$ whatever the value $X$ has.

## 2.4.1  Distributive Theorem

$$X(Y + Z) = XY + XZ$$

**Table 2.9** The truth table for $X(Y + Z) = XY + XZ$

| X | Y | Z | Y + Z | X(Y+Z) | XY | XZ | XY + XZ |
|---|---|---|-------|--------|----|----|---------|
| 0 | 0 | 0 | 0 | 0 | 0 | 0 | 0 |
| 0 | 0 | 1 | 1 | 0 | 0 | 0 | 0 |
| 0 | 1 | 0 | 1 | 0 | 0 | 0 | 0 |
| 0 | 1 | 1 | 1 | 0 | 0 | 0 | 0 |
| 1 | 0 | 0 | 0 | 0 | 0 | 0 | 0 |
| 1 | 0 | 1 | 1 | 1 | 0 | 1 | 1 |
| 1 | 1 | 0 | 1 | 1 | 1 | 0 | 1 |
| 1 | 1 | 1 | 1 | 1 | 1 | 1 | 1 |

In order to prove the above theorem, the truth table of both sides of the theorem is generated as shown in Table 2.9 and shows both sides generating the same truth table.

### 2.4.2 De Morgan's Theorem I

$$(X + Y)' = X' \ Y'$$

In this theorem the OR between X and Y is negated and changes the OR operation to the AND operation.

*Proof*: By making a truth table for both sides of the theorem, it shows that both sides of the theorem generate the same truth table (Table 2.10).

### 2.4.3 De Morgan's Theorem II

$$(XY)' = X' + Y'$$

In this theorem the XY is complemented and changes it from an AND operation to an OR operation with each component complemented.

*Example*: $(WXYZ)' = W' + X' + Y' + Z'$

If the truth table of both sides were generated, then it would show that both sides have the same truth table result.

### 2.4.4 Commutative Law

**Table 2.10** The truth table showing De Morgan's Law

| X | Y | X + Y | (X + Y)' | X' | Y' | X' Y' |
|---|---|-------|----------|----|----|-------|
| 0 | 0 | 0 | 1 | 1 | 1 | 1 |
| 0 | 1 | 1 | 0 | 1 | 0 | 0 |
| 1 | 0 | 1 | 0 | 0 | 1 | 0 |
| 1 | 1 | 1 | 0 | 0 | 0 | 0 |

$$X + Y = Y + X$$
$$XY = YX$$

### 2.4.5  Associative Law

$$X(YZ) = (XY)Z$$
$$X + (Y + Z) = (X + Y) + Z$$

### 2.4.6  More Theorems

The following are useful theorems:

(a) $X + X'Y = X + Y$
(b) $X' + XY = X' + Y$
(c) $X + X'Y' = X + Y'$
(d) $X' + XY' = X' + Y'$

**Example: Simplify the Following Functions**
(a) $F(X, Y, Z) = XY'Z + XY'Z' + XY$
$\quad F(X, Y, Z) = XY'(Z + Z') + YZ$ where $Z + Z' = 1$ then
$\quad F(X, Y, Z) = XY' + XY = X(Y + Y') = X$
(b) $F(X, Y, Z) = (X' + Y)(X + Y') = X'X + X'Y' + XY + YY'$ where $X'X$ and $YY'$
are zero then $F(X, Y, Z) = X'Y' + XY$

## 2.5  Boolean Function

Boolean function is represented by algebraic expression which is made of binary variables such as X, Y, and Z and logic operations between variables such as AND, OR, and NOT.

$F(X, Y, Z) = X + YZ$ is a Boolean function.

Figure 2.19 shows the logic circuit for function F where X, Y, and Z are the inputs and F is the output; Table 2.11 shows the truth table of the function F.

**Fig. 2.19** Logic circuit for a function F(X, Y, Z) = X + YZ

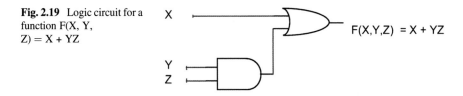

$$F(X,Y,Z) = X + YZ$$

**Table 2.11** The truth table for function F(X, Y, Z) = X + YZ

| X | Y | Z | YZ | X + YZ |
|---|---|---|----|--------|
| 0 | 0 | 0 | 0  | 0      |
| 0 | 0 | 1 | 0  | 0      |
| 0 | 1 | 0 | 0  | 0      |
| 0 | 1 | 1 | 1  | 1      |
| 1 | 0 | 0 | 0  | 1      |
| 1 | 0 | 1 | 0  | 1      |
| 1 | 1 | 0 | 0  | 1      |
| 1 | 1 | 1 | 1  | 1      |

The truth table shows the characteristics of function F; the function F = 1 when the inputs to the circuit are 100 or 101 or 110 or 111.

## 2.5.1   Complement of a Function

In order to complement a function, both sides of the function must be complemented.
*Example*: Complement the following function:

$$F(X, Y, Z) = XY + Y'Z$$

$F'(X, Y, Z) = (XY + Y'Z)'$ using De Morgan's theorem.
$F'(X, Y, Z) = (XY)'(Y'Z)'$
$F'(X, Y, Z) = (X' + Y')(Y + Z')$

*Example*: Find the complement of the following function:

$$F(X, Y, Z) = (X' + Y')(Y + Z')$$

Complement both sides of the function:

$$F'(X, Y, Z) = [(X' + Y')(Y + Z')]'$$

Applying De Morgan's theorem results

$$F'(X, Y, Z) = [(X' + Y')]' + [(Y + Z')]'$$
$$F'(X, Y, Z) = [(XY)] + [(Y'Z)]$$
$$F'(X, Y, Z) = (XY) + (Y'Z)$$

## 2.6  Summary

- Boolean logic consists of the AND, OR, and NOT logics.
- The output of the 2-input AND gate is one when both inputs are one; otherwise the output is zero.
- The output of the 2-input OR gate is one when at least one of the inputs is one; otherwise the output is zero.
- NOT gate performs one's complement.
- The integrated circuits (IC) are classified by SSI (small-scale integration), MSI (medium-scale integration), LSI (large-scale integration), and very-large-scale integration.
- A NAND gate is equivalent of the AND-NOT.
- A NOR gate is equivalent of the NOR-NOT.
- The 2-input XOR has output one when the inputs are not equal; otherwise the output is zero.
- The 2-input exclusive NOR (XNOR) is equivalent of the XOR-NOT.
- Chapter 3 will cover minterms and maxterms, apply K-map to simplify a function, and use universal gates to draw logic circuit.

## Problems

1. Show truth table for 4-input AND, OR, NOR, and NAND gates.
2. If A = 11001011 and B = 10101110, then what is the value of the following operations:

   (a) A AND B
   (b) A OR B

3. If A = 11001011 and B = 10101110, what is the value of the following operations (F in hex = 1111):

   (a) A NOT
   (b) A XOR B
   (c) A AND 0F
   (d) A AND F0

4. Draw a logic circuit for the following functions:

   (a) $F(X, Y, Z) = XY' + XZ' + YZ$
   (b) $F(X, Y, Z) = (X + Y') (X' + Z')(Y + Z)$

5. Boolean theorems to simplify the following expressions:

   (a) $X + X + X$
   (b) $XY + XY$
   (c) $YYY$
   (d) $X + XY$
   (e) $XY' + Y'$
   (f) $(X + Y)Y'$
   (g) $(XY) + (XY)'$
   (h) $X'Y' + XY$

6. Simplify the following functions:

   (a) $F(X, Y, Z) = XY + X'Y + XZ$
   (b) $F(X, Y, Z) = (X + Y) (X' + Y + Z)$
   (c) $F(X, Y, Z) = XY'Z + XYZ + Y'Z$
   (d) $F(X, Y, Z) = XY + X'YZ$
   (e) $F(X, Y, Z) = X'Y + XYZ'$
   (f) $F(X, Y, Z) = (XY) + (X + Y + Z)'X + YZ$
   (g) $F(X, Y, Z) = (XY)' + (X + Y + Z)'$

7. Find the truth table for the following functions:

   (a) $F(X, Y, Z) = XY' + YZ + XZ'$.
   (b) $F(X, Y, Z) = (X + Y')(Y + Z)(X' + Z')$

8. If $A = 10110110$ and $B = 10110011$, then find

   (a) A NAND B
   (b) A NOR B
   (c) A XOR B

9. Find the output of the following gates:

   (a)

   (b)

   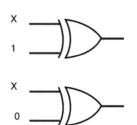

   (c)

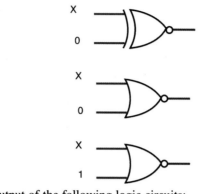

(d)

(e)

10. Show the output of the following logic circuits:

(a)

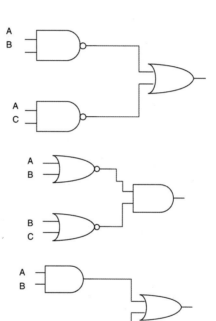

(b)

(c)

11. Find the output function of the following logic circuits:

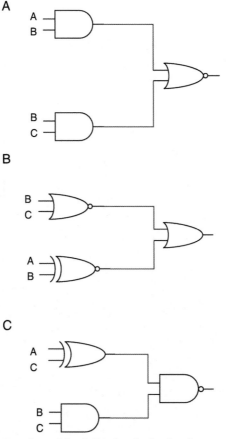

12. Find the output function of the following logic circuit:

13. Draw a logic circuit and show the truth table for the following functions:

    (a) $F(X, Y) = (XY)' + X(X + Y')$
    (b) $F(X, Y, Z) = (X + Y + Z')' (X' + Y')$
    (c) $F(X, Y, Z) = (X \text{ XOR } Y) (X \text{ NOR } Y')$
    (d) $F(X, Y, Z) = (X' + Y' + Z) (X + Y)$

14. Show truth table for each of the following functions:

    (a) $F(X, Y, Z) = XY' + XZ' + YZ$
    (b) $F(X, Y, Z) = (X + Y) (X + Z')$
    (c) $F(X, Y, Z) = XY (Y + Z')$
    (d) $F(X, Y, Z) = (X + Y)'(X' + Z)$

15. Simplify the following functions:

    (a) $F(X, Y, Z) = YZ + (X + Y)' + (XYZ)'$.
    (b) $F(X, Y, Z) = (X + Y + Z)' (X + Y)$

16. Draw logic circuits for the following functions:

    (a) $F(X, Y, Z) = (X + Y)' + YZ$
    (b) $F(X, Y, Z) = (XYZ)' + XZ + YZ$

# Chapter 3
# Minterms, Maxterms, Karnaugh Map (K-Map), and Universal Gates

**Objectives: After Completing this Chapter, you Should Be Able to**
- Represent a Boolean function in the form of sum of minters and product of maxterm.
- Generate a truth table from a function that is represented by the sum of minterms.
- Generate a truth table from a function that is represented by the product of minterms.
- Developed a function from the truth table.
- Use K-map to simplify a function.
- Apply don't care condition in a K-map.
- Draw logic circuit using only NAND or NOR gates.

## 3.1 Introduction

Digital circuit can be represented by a truth table or Boolean function. In a digital circuit with multiple digital inputs and multiple digital outputs, the outputs depend on the current value of inputs. A Boolean function can be represented in the form of *sum of minterms* or the *product of maxterms*, which enable the designer to make a truth table more easily. Also, Boolean functions can be simplified using Karnaugh **map (K-map)** without using Boolean theorems, by transferring a function to K-map and reading simplified function from K-map. Most digital systems are designed by using universal gates (NAND or NOR).

© The Author(s), under exclusive license to Springer Nature Switzerland AG 2022
A. Elahi, *Computer Systems*, https://doi.org/10.1007/978-3-030-93449-1_3

## 3.2 Minterms

A *minterm* is associated with each combination of variables in a function. If a function has $n$ variables, then it has $2^n$ minterms. Consider two Boolean variables, X and Y. There are four different possible combinations that can be generated from X AND Y, and they are $\overline{X}\overline{Y}$, $\overline{X}Y$, $X\overline{Y}$, and XY. These four combinations are called the *minterms* for X AND Y. Table 3.1 shows the minterms and their designations for F(X,Y) = (X AND Y).

In Table 3.1, $\overline{X}\,\overline{Y} = 1$, if X = 0 and Y = 0, then $\overline{X}\,\overline{Y}$ is represented by $m_0$ (decimal number of 00); $\overline{X}Y = 1$, if X = 0 and Y = 1, then $\overline{X}Y$ is represented by $m_1$; $X\overline{Y} = 1$, if X = 1 and Y = 0 and $X\overline{Y}$ is represented by $m_2$; XY = 1, if X = 1 and Y = 1, then XY is represented by $m_3$.

### 3.2.1 Application of Minterms

It is simple to generate a truth table from minterms and vice versa. Consider the function $F(X,Y) = X\overline{Y} + \overline{X}Y$ and its truth table (Table 3.2), this function can be represented as $F(X,Y) = m_1 + m_2$ or each minterm that represents a *one* in the truth table. This may also be rewritten as $F(X,Y) = \sum(1, 2)$.

### 3.2.2 Three-Variable Minterms

The three variables X, Y, and Z generate eight minterms as shown in Table 3.3.

**Table 3.1** Minterms of F(X,Y)

| X Y | Minterm | Designation |
|-----|---------|-------------|
| 0 0 | $\overline{X}\overline{Y}$ | $m_0$ |
| 0 1 | $\overline{X}\,Y$ | $m_1$ |
| 1 0 | $X\,\overline{Y}$ | $m_2$ |
| 1 1 | $X\,Y$ | $m_3$ |

**Table 3.2** Truth table for function $F(X,Y) = X\overline{Y} + \overline{X}Y$ with minterms

| X | Y | F | |
|---|---|---|---|
| 0 | 0 | 0 | $m_0$ |
| 0 | 1 | 1 | $m_1$ |
| 1 | 0 | 1 | $m_2$ |
| 1 | 1 | 0 | $m_3$ |

**Table 3.3** Three-variable minterms

| X Y Z | Minterms | Designation |
|-------|----------|-------------|
| 0 0 0 | X'Y'Z' | $m_0$ |
| 0 0 1 | X'Y'Z | $m_1$ |
| 0 1 0 | X'YZ' | $m_2$ |
| 0 1 1 | X'YZ | $m_3$ |
| 1 0 0 | XY'Z' | $m_4$ |
| 1 0 1 | XY'Z | $m_5$ |
| 1 1 0 | XYZ' | $m_6$ |
| 1 1 1 | XYZ | $m_7$ |

**Table 3.4** Truth table for function F(X,Y, Z) = X'Y'Z + X'YZ + XYZ

| X Y Z | F |
|-------|---|
| 0 0 0 | 0 |
| 0 0 1 | 1 |
| 0 1 0 | 0 |
| 0 1 1 | 1 |
| 1 0 0 | 0 |
| 1 0 1 | 0 |
| 1 1 0 | 0 |
| 1 1 1 | 1 |

**Example 3.1** Find the truth table for the following function:

$$F(X, Y, Z) = X'Y'Z + X'YZ + XYZ$$

The function F can be represented by a *sum of the minterms* (or where F = 1):

$$F(X, Y, Z) = m_1 + m_3 + m_7$$

or

$$F(X, Y, Z) = \sum(1, 3, 7)$$

The truth table for this function contains a *one* in row 1, row 3, and row 7. The rest of the rows are *zeros* as shown in Table 3.4. The function for a truth table can also be determined from the sum of the minterms.

**Example 3.2** The following truth table is given; find the function F.

In the following table, the output of function F is one when the input is $001 = m_1$, $011 = m_3$, $101 = m_5$, and $111 = m_7$; therefore $F(X,Y,Z) = m_1 + m_3 + m_5 + m_7$ or

$$F(X, Y, Z) = \sum(1, 3, 5, 7)$$

X Y Z   F
0 0 0    0

$$
\begin{array}{ccc}
0\ 0\ 1 & 1 \\
0\ 1\ 0 & 0 \\
0\ 1\ 1 & 1 \\
1\ 0\ 0 & 0 \\
1\ 0\ 1 & 1 \\
1\ 1\ 0 & 0 \\
1\ 1\ 1 & 1
\end{array}
$$

Substituting each designation of a minterm with its actual product term (e.g., $m_0 = X'Y'Z'$) results in the following function:

$$F(X, Y, Z) = X'Y'Z + X'YZ + XY'Z + XYZ$$

**Example 3.3**  For the following truth table:

(a) Find the function F.
(b) Simplify the function.
(c) Draw the logic circuit for the simplified function.

$$
\begin{array}{ccc}
X\ Y\ Z & F \\
0\ 0\ 0 & 1 \\
0\ 0\ 1 & 0 \\
0\ 1\ 0 & 1 \\
0\ 1\ 1 & 1 \\
1\ 0\ 0 & 0 \\
1\ 0\ 1 & 0 \\
1\ 1\ 0 & 0 \\
1\ 1\ 1 & 1
\end{array}
$$

From the truth table, where $F = 1$, we select the minterms $m_0$, $m_2$, $m_3$, and $m_7$:

$$F(X, Y, Z) = m_0 + m_2 + m_3 + m_7$$

Or

$$F(X, Y, Z) = \overline{X}\,\overline{Y}\,\overline{Z} + \overline{X}Y\overline{Z} + \overline{X}YZ + XYZ$$
$$\overline{XZ} \qquad \overline{Z}$$

$$F(X, Y, Z) = \overline{XZ}\ (\overline{Y} + Y) + YZ(\overline{X} + X)$$
$$F(X, Y, Z) = \overline{XZ} + YZ$$

The logic circuit for the simplified function F is given in Fig. 3.1.

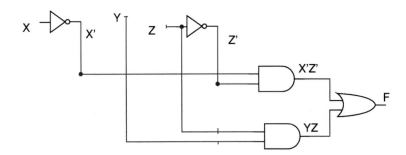

**Fig. 3.1** Logic circuit for $F(X,Y,Z) = \overline{X}\,\overline{Z} + YZ$

**Table 3.5** Maxterms of function $F(X,Y,Z)$

| X Y Z | Maxterm | Designation |
|---|---|---|
| 0 0 0 | X + Y + Z | $M_0$ |
| 0 0 1 | X + Y + Z' | $M_1$ |
| 0 1 0 | X + Y' + Z | $M_2$ |
| 0 1 1 | X + Y' + Z' | $M_3$ |
| 1 0 0 | X' + Y + Z | $M_4$ |
| 1 0 1 | X' + Y + Z' | $M_5$ |
| 1 1 0 | X' + Y' + Z | $M_6$ |
| 1 1 1 | X' + Y' + Z' | $M_7$ |

**Table 3.6** Truth Table for $F(X,Y,Z) = M_0 M_2 M_4 M_5 M_6$

| X Y Z | F |
|---|---|
| 0 0 0 | 0 |
| 0 0 1 | 1 |
| 0 1 0 | 0 |
| 0 1 1 | 1 |
| 1 0 0 | 0 |
| 1 0 1 | 0 |
| 1 1 0 | 0 |
| 1 1 1 | 1 |

## 3.3 Maxterms

Maxterm is complement of a minterm. If the *minterm* $\mathbf{m_0}$ is $(\overline{XYZ})$, then the *maxterm* $\mathbf{M_0}$ is

$$\left(\overline{\overline{XYZ}}\right) = X + Y + Z$$

Table 3.5 shows the maxterms for three variables.

In a truth table, an output of *one* represents *minterms*, and an output of *zero* represents *maxterms*. Consider the truth Table 3.6, where the function F can be expressed as the *product of maxterms*. (Please note: the *product* of maxterms, as opposed to the *sum* of minterms.)

$F(X,Y,Z) = M_0M_2M_4M_5M_6$, or it can be represented by

$$F(X, Y, Z) = \pi(0, 2, 4, 5, 6)$$

Substituting each designated maxterm with the corresponding maxterm results in:

$$F(X, Y, Z) = (X + Y + Z)(X + Y' + Z)(X' + Y + Z)(X' + Y + Z')$$

## 3.4   Karnaugh Map (K-Map)

Karnaugh maps are used to simplify a Boolean function without using Boolean algebra theorems. A K-map is also another way to represent the truth table of a function. K-maps are made of cells where each cell represents a minterm. Cells marked with a *one* will be the minterms used for the *sum of the minterms* representation of a function. Conversely, cells marked with a *zero* will be used for the *product of the maxterms* representation.

Two variables X and Y can have four minterms as shown in Table 3.7. Each minterm is represented by a cell in the K-map, so a two-variable K-map contains four cells as shown in Fig. 3.2.

In Fig. 3.2 each cell represents a minterm. The cell located at row 0 and column 0 represents $m_0$ (minterm zero) or $X'Y'$. The cell located at row 1 and column 1 is represented by $m_3$ or XY. As shown in Fig. 3.2, both cells in row zero contain $X'$, so this row is labeled the $X'$ row. Both cells in row 1 contain X so that row is labeled as the X row.

**Table 3.7** Minterms for two variables

| X Y | Minterms | Designation |
|-----|----------|-------------|
| 0 0 | X'Y' | $m_0$ |
| 0 1 | X'Y | $m_1$ |
| 1 0 | XY' | $m_2$ |
| 1 1 | XY | $m_3$ |

**Fig. 3.2** Two-variable K-map

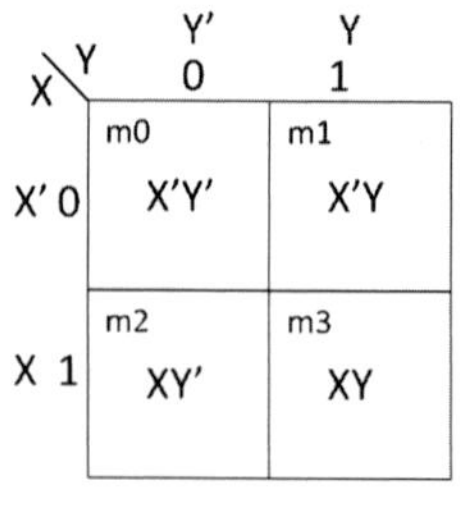

**Fig. 3.3** The function in a K-map

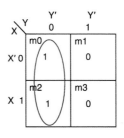

**Table 3.8** The truth table for the function F(X, Y) = XY' + X'Y'

| X Y | F |
|-----|---|
| 0 0 | 1 |
| 0 1 | 0 |
| 1 0 | 1 |
| 1 1 | 0 |

A K-map of a function is another way to represent the truth table of the function, as seen in Fig. 3.3.

Consider the function $F(X,Y) = XY' + X'Y' = m_2 + m_0$. The truth table for the function is given in Table 3.8, and the function is also mapped to a K-map as shown in Fig. 3.3.

**Adjacent Cell**

Two cells are adjacent if they differ on only one variable. The cells $X'Y'$ and $X'Y$ are adjacent because their only difference is $Y'$ and $Y$. Adjacent cells can be combined in order to simplify a K-map's function.

As shown in the K-map, the cells $m_0$ and $m_2$ contain ones and the other cells contain zeros. The cells $m_0$ and $m_2$ are adjacent to each other. Note that the adjacent cells take up the entire column of $Y'$, and all other cells are *zero*. Our simplified function is therefore $F(X,Y) = Y'$.

**Example 3.4** Simplify the following function:

$$F(X, Y) = X'Y + XY' + XY$$

or

$$F(X, Y) = m_1 + m_2 + m_3$$

Transferring minterms into a K-map results in Fig. 3.4.

As shown in Fig. 3.4, the cells $m_2$ and $m_3$ are adjacent, so they can be combined. Likewise, the cells $m_1$ and $m_3$ can be combined. By reading the map, you will have the simplified function.

Cells $m_2$ and $m_3$ are the entire row X, and cells $m_1$ and $m_3$ are the entire column Y, with the other cell being *zero*. Therefore,

**Fig. 3.4** K-map for F(X, Y) = X'Y + XY' + XY

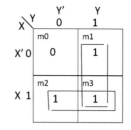

**Fig. 3.5** Three-variable Map

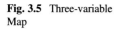

$$F(X, Y) = X + Y$$

### 3.4.1   Three-Variable Map

A three-variable K-map contains eight cells, and each cell represents a minterm as shown in Fig. 3.5.

Observe the following about the K-map in Fig. 3.5:

(a) At row 0, all four cells contain X'; therefore this row is labeled X'.
(b) At row 1, all four cells contain X; therefore this row is labeled X.
(c) At the columns 11 and 10, all four cells contain Y; therefore these columns are labeled Y.
(d) At the columns 00 and 01, all four cells contain Y'; therefore these columns are labeled Y'.
(e) At the columns 01 and 11, all four cells contain Z; therefore these columns are labeled Z.
(f) At the columns 00 and 10, all four cells contain Z'; therefore these columns are labeled Z'.

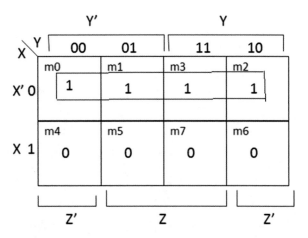

**Fig. 3.6** The grouping of four cells in a K-map where the simplified F(X,Y, Z) = X′

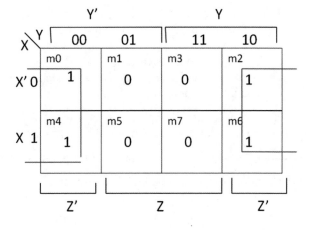

**Fig. 3.7** Combing the four Z′ cells together in the K-map

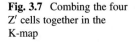

Adjacent cells can be grouped together in a K-map; in a K-map it can combine 2 cells, 4 cells, 8 cells, and 16 cells. Figure 3.6 shows how the *ones* could be grouped in a K-map.

In Fig. 3.6, all four cells can be combined. Folding the K-map horizontally twice will result in all of the *ones* overlapping, and row X′ covers all four *ones*.

For Fig. 3.7, consider folding a K-map once more. The four *ones* will overlap if the map is folded once horizontally, then vertically. Also note that Z′ covers all four *ones*.

**Example 3.5** Simplify the following function:

$$F(X, Y, Z) = X'Y + XZ + XZ'$$

First, each term of the function must be transferred to the K-map.

**Fig. 3.8** The combination of the cells of the example function in a K-map

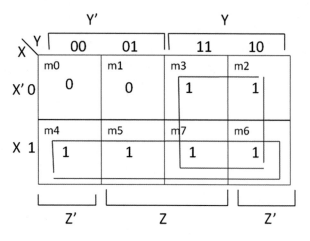

(a) The first term being X'Y, place a *one* on each cell located at the intersection of the X' row and the Y column as shown in Fig. 3.8. ($m_3$ and $m_2$).

(b) The second term is XZ, so place a *one* on each cell located at the intersection of the Y' column and the X row. ($m_5$ and $m_7$).

(c) The third term being XZ', place a *one* on each cell in the intersection of the X row and the Z column. ($m_4$ and $m_6$).

Since our adjacent cells include the entire row X and every cell in the columns Y, a simplified form of this function would be $F(X,Y,Z) = X + Y$.

**Example 3.6** Read the K-maps of Fig. 3.9a–d to determine the simplified function.

(a) All adjacent cells in columns Z' and columns Y are *one*.

$$F(X, Y, Z) = Z' + Y$$

(b) The cells in row X', columns Y' are adjacent, as are the cells in row X, columns Y.

$$F(X, Y, Z) = X'Y' + XY$$

(c) All cells are *one*, so the function is always equal to *one*.

$$F(X, Y, Z) = 1$$

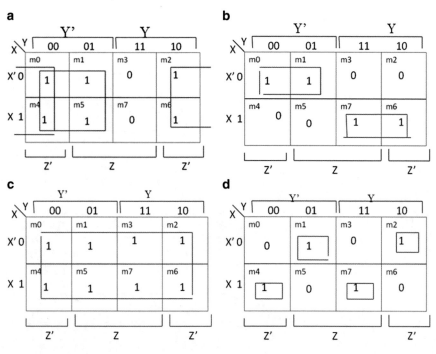

**Fig. 3.9** K-Maps for Example 3.6

(d) Without adjacent cells to simplify the terms, the function equals the *ones*:

$$F(X, Y, Z) = X'Y'Z + X'YZ' + XY'Z' + XYZ$$

### 3.4.2 Four-Variable K-Map

Four-variable K-maps contain 16 cells as shown in Fig. 3.10. Please note the specific layout of the map.

The following describes the coverage of each variable by K-map:

- W covers rows 11 and 10.
- W' covers rows 00 and 01.
- X covers rows 01 and 11.
- X' covers rows 00 and 10.
- Y covers columns 11 and 10.
- Y' covers columns 00 and 01.
- Z covers columns 01 and 11.
- Z' covers columns 00 and 10.

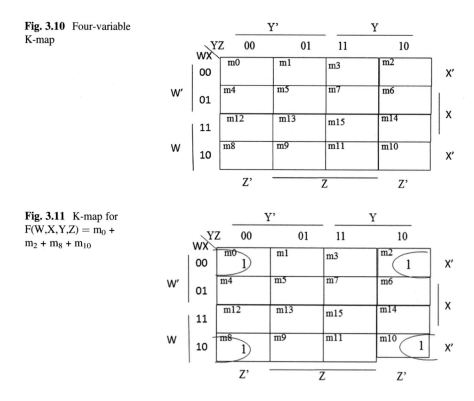

**Fig. 3.10** Four-variable K-map

**Fig. 3.11** K-map for $F(W,X,Y,Z) = m_0 + m_2 + m_8 + m_{10}$

**Example 3.7** Simplify the following function:

$$F(W, X, Y, Z) = m_0 + m_2 + m_8 + m_{10}.$$

The function is transferred to the K-map as shown in Fig. 3.11. If the K-map is folded once vertically and horizontally from the middle, then all four cells containing *one* overlap each other. Note that each of these cells makes up all intersections of $X'$ and $Z'$.

The simplified function is $F(W,X,Y,Z) = X'Z'$.

**Example 3.8** Read the following K-map:

$$F(W, X, Y, Z) = X'Y' + X'Z + XYZ'$$

## 3.5    Sum of Products (SOP) and Product of Sums (POS)

The *sum of products* of a function is its simplified *sum of minterms*.

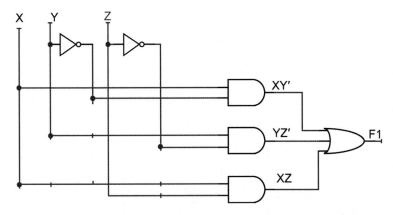

**Fig. 3.12**   Logic circuit for F1(X,Y,Z) = XY′ + YZ′ + XZ made of AND-OR gates

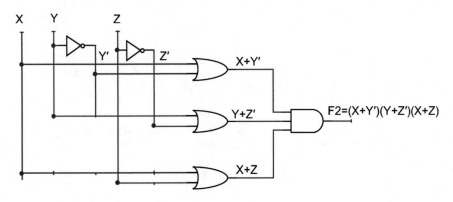

**Fig. 3.13**   Logic circuit for F2(X,Y,Z) = (X + Y′)(Y + Z′)(X + Z)

Observe the function F(X,Y) = XY′ + XY is in the form of SOP where the addition sign is used for OR logic, and XY is called product. Consider function F1 (X,Y,Z) = XY′ + YZ′ + XZ, and Fig. 3.12 shows logic circuit for function F1 which is made of AND-OR.

Consider function F2(X,Y,Z) = (X + Y′)(Y + Z′)(X + Z) which is represented by the product of sums, and Fig. 3.13 shows logic circuit for function F2 which is in the form OR-AND.

**Example 3.9** Simplify the following function in the form of SOP and POS.

- **F(X,Y,Z) = $\Sigma$(0,1,6,7)**

Combining the *ones* (the minterms $m_0$, $m_1$, $m_6$, and $m_7$) in a K-map results in Fig. 3.14.

The *sum of products* is therefore: F(X,Y,Z) = X′Y′ + XY.

Combining the *zeros* in a K-map returns the significant *maxterms* as in Fig. 3.15. If F equals $\Sigma$(0,1,6,7), then F′ equals $\pi(M_2 M_3 M_4 M_5)$.

**Fig. 3.14** K-map for F(X,
Y,Z) = $\sum(0,1,6,7)$

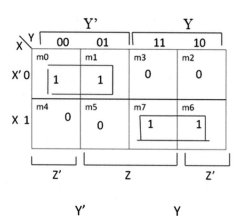

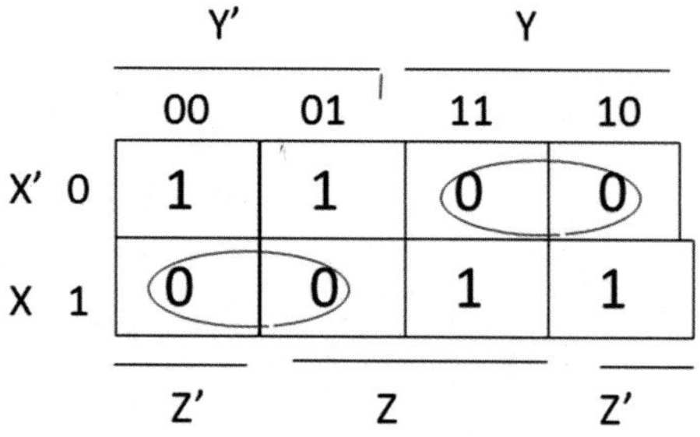

**Fig. 3.15** K-map for F'(X,
Y,Z) = $\pi(M_2M_3M_4M_5)$

Since the *product of maxterms* is equal to the *complement* of F, in order to find F, both sides of the function will be complemented:

$$F(X, Y, Z) = (XY' + X'Y)'$$

Using De Morgan's theorem

$$F(X, Y, Z) = (XY')'(X'Y)'$$

or
F(X,Y,Z) = (X' + Y)(X + Y') in the final, *product of sums* form.

## 3.6  *Don't Care* Conditions

In a truth table, if certain combinations of the input variables are impossible, they are considered *don't care* conditions. These conditions are where the output of the function does not matter. For example, binary-coded decimal (BCD) is 4 bits and only 0000 to 1001 are used, so from 1010 to 1111 are not BCD; the truth table or K-map values are *don't cares*. A truth table or K-map cell marked with a "X" or "d" is a *don't care* term, and output will not be affected whether it is a *one* or *zero*. The

**Fig. 3.16** K-map with *don't care* minterms

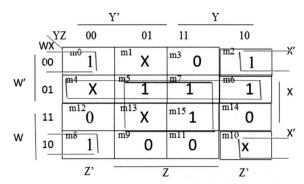

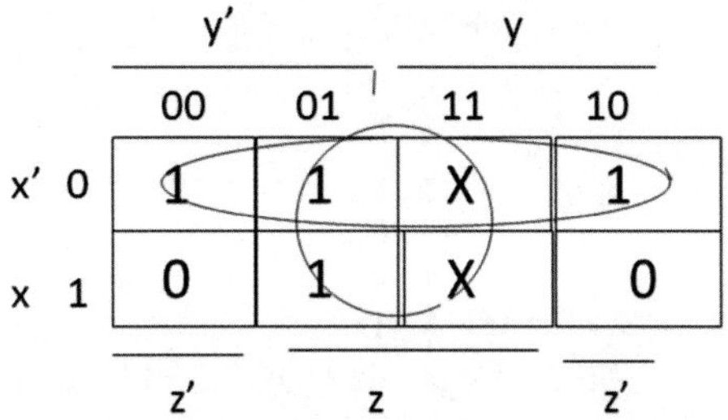

**Fig. 3.17** K-map for F(X, Y,Z) = $m_0 + m_1 + m_2 + m_5$ and D(X,Y,Z) = $m_3 + m_7$

*don't care* can be used to expand the adjacency of cells in a K-map to further simplify a function, since their output does not matter.

**Example 3.10** Figure 3.16 shows a K-map with *don't care* minterms at $m_1$, $m_{10}$, and $m_{13}$. Since a *don't care* can output either a *zero* or *one*, we can assume it is a *one* in order to expand a grouping of adjacent cells.

From Fig. 3.16, the function would be $F(W,X,Y,Z) = XZ + X'Z' + XW'$.

When minterms for function F are *don't care* terms, the *don't care function* D is equal to the sum of the *don't care* minterm(s). If $m_7$ is the only *don't care*, then don't care function is represented by $D(X,Y,Z) = m_7$.

**Example 3.11** Simplify the following function where D is a *don't care* function:

$$F(X, Y, Z) = m_0 + m_1 + m_2 + m_5$$
$$D(X, Y, Z) = m_3 + m_7$$

Using these values results in the K-map in Fig. 3.17. By grouping adjacent cells and using the *don't care* terms, $F(X,Y,Z) = X' + Z$.

## 3.7  Universal Gates

The NAND and NOR gates are called universal gates. With NANDs or NORs, designers are able to construct other logic gates such as OR, AND, and NOT gates.

### 3.7.1  Using NAND Gates

(a) **NOT from NAND**

A NOT gate is generated by connecting the inputs of a NAND gate together as shown in the following figure.

Or, it can be represented by:

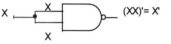

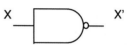

(b) **AND from NAND**

An AND gate is constructed by connecting the inputs to a NAND gate and putting another NAND on the output (to act as a NOT gate).

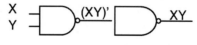

(c) **OR from NAND**

An OR gate is constructed by connecting each input to an individual NAND gate and putting every output into a single NAND which acts as a NOT gate.

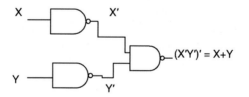

### 3.7.2  Using NOR Gates

(a) **NOT from NOR**

A NOT gate is generated by connecting the inputs of a NOR gate together as shown in the following figure.

Or, it can be represented by:

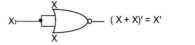

**(b) OR from NOR**

An OR gate is constructed by connecting the inputs to a NOR gate and putting another NOR on the output (to act as a NOT gate).

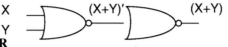

**(c) AND from NOR**

An AND gate is constructed by connecting each input to an individual NOR gate and putting every output into a single NOR which acts as a NOT gate.

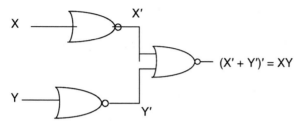

### 3.7.3   Implementation of Logic Functions Using NAND Gates or NOR Gates Only

Many logic functions are implemented using only NAND or NOR gates, rather than a combination of various gates. Most logic gate ICs contain multiple gates of a single type, such as an IC containing eight AND gates. Using a single type of gate can reduce the number of ICs needed.

Consider the function $F(X,Y) = X'Y' + XY$ and its logic circuit diagram in Fig. 3.18.

This diagram would require one IC for AND gates, another for NOT gates, and one OR gate, or a total of three separate ICs.

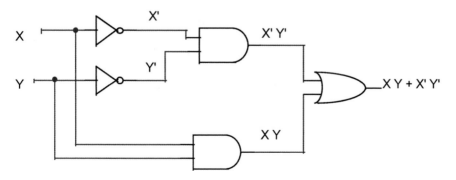

**Fig. 3.18**  Logic circuit for $F(X,Y) = XY + X'Z'$

By complementing the function twice, the right side of the equation may be easier to use with NAND and NOR gates.

**Example 3.12** Create a logic circuit using only NAND and only NOR for the function $F(X,Y) = X'Y' + XY$.

### 3.7.4 Using NAND Gates

- Complement the right side of the equation twice.
- $F(X,Y) = X'Y' + XY \rightarrow F(X, Y) = [(X'Y' + XY)']'$.
- Use Boolean theorems to make it NAND friendly:

  - $F(X,Y) = [(X'Y')'(XY)']'$

Consider the final function once more: $F(X, Y) = [(X'Y')'(XY)']'$, and substitute placeholders for the inner terms (Fig. 3.19).

- $F = [(X'Y')'(XY)']' = [AB]'$ (A *NAND* B)

  - $A = (X'Y')' = X'$ *NAND* $Y'$
  - $B = (XY)' = X$ *NAND* $Y$
  - $X' = X$ *NAND* $X$
  - $Y' = Y$ *NAND* $Y$

### 3.7.5 Using NOR Gates

- Complement the right side of the equation twice:

  - $F(X,Y) = X'Y' + XY \rightarrow F(X, Y) = [(X'Y' + XY)']'$

- Use Boolean theorems to make it NOR friendly:

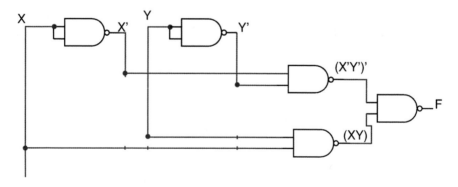

**Fig. 3.19** Logic circuit of $F(X,Y) = X'Y' + XY$ using only NAND gates

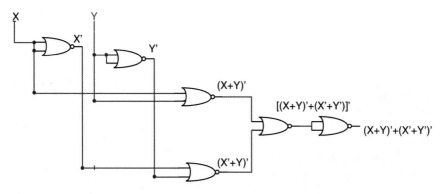

**Fig. 3.20** Logic circuit of $F(X,Y) = X'Y' + XY$ using only NOR gates

- $F(X,Y) = [(X'Y')'(XY)']'$
- $F(X,Y) = [(X + Y)(X' + Y')]'$
- $F(X,Y) = [(X + Y)' + (X' + Y')']'$

Consider the final function once more: $F(X, Y) = [(X + Y)' + (X' + Y')']'$, and substitute placeholders for the inner terms (Fig. 3.20).

- $F = [(X'Y')'(XY)']' = [A + B]'$ (A *NOR* B)

  - $A = (X + Y)' = X$ *NOR* $Y$
  - $B = (X' + Y')' = X'$ *NOR* $Y'$
  - $X' = X$ *NOR* $X$
  - $Y' = Y$ *NOR* $Y$

## 3.8 Summary

- A digital circuit is made of the combination of the different gates with the multiple digital inputs and the multiple digital outputs; the outputs depend only to the current values of the inputs.
- A combinational logic circuit can be represented by a Boolean function or truth table.
- Boolean theorems or a K-map can be used to simplify a Boolean function.
- A Boolean function can be represented by the sum of the products (SOP) or the product of sums (POS).
- The NAND and NOR gates are called universal gates. It can generate other gates by using NAND or NOR gates.
- Don't care condition is the input value that never applied to a combinational circuit which results output with don't care condition (0 or 1).
- Chapters 1, 2, and 3 cover basic topics in order to be able to design digital system; Chap. 4 presents how to design a digital system and covers digital components

that are used for designing digital system such as decoder, multiplexer, binary adder, binary subtractor, and arithmetic logic unit (ALU).

## Problems

1. Find the output function of each truth table:

   (a) As the sum of minterms
   (b) As the product of maxterms

| X Y Z | F |
|-------|---|
| 0 0 0 | 1 |
| 0 0 1 | 0 |
| 0 1 0 | 0 |
| 0 1 1 | 1 |
| 1 0 0 | 0 |
| 1 0 1 | 1 |
| 1 1 0 | 0 |
| 1 1 1 | 1 |

| A B C D | F |
|---------|---|
| 0 0 0 0 | 1 |
| 0 0 0 1 | 0 |
| 0 0 1 0 | 1 |
| 0 0 1 1 | 1 |
| 0 1 0 0 | 0 |
| 0 1 0 1 | 1 |
| 0 1 1 0 | 1 |
| 0 1 1 1 | 1 |
| 1 0 0 0 | 1 |
| 1 0 0 1 | 1 |
| 1 0 1 0 | 0 |
| 1 0 1 1 | 0 |
| 1 1 0 0 | 0 |
| 1 1 0 1 | 1 |
| 1 1 1 0 | 1 |
| 1 1 1 1 | 1 |

2. Generate truth table for the following functions:

   (a) $F(X,Y,Z) = \Sigma(1,3,6,7)$
   (b) $F(X,Y,Z) = \pi(1,3,4)$
   (c) $F(W,X,Y,Z) = \Sigma(1,4,7,10,12,15)$
   (d) $F(W,X,Y,Z) = \pi(2,3,4,7,10,11,12,13)$

3. Generate the function F for the following K-maps:

(a)

| X\YZ | Y' 00 | 01 | Y 11 | 10 |
|---|---|---|---|---|
| X' 0 | m0 **1** | m1 **0** | m3 **1** | m2 **1** |
| X 1 | m4 **1** | m5 **1** | m7 **0** | m6 **1** |

Z' | Z | Z'

(b)

| X\YZ | Y' 00 | 01 | Y 11 | 10 |
|---|---|---|---|---|
| X' 0 | m0 **0** | m1 **0** | m3 **1** | m2 **1** |
| X 1 | m4 **1** | m5 **1** | m7 **1** | m6 **1** |

Z' | Z | Z'

(c)

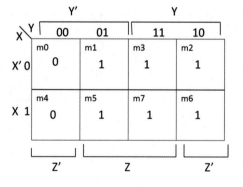

| X\YZ | Y' 00 | 01 | Y 11 | 10 |
|---|---|---|---|---|
| X' 0 | m0 **0** | m1 **1** | m3 **1** | m2 **1** |
| X 1 | m4 **0** | m5 **1** | m7 **1** | m6 **1** |

Z' | Z | Z'

(d)

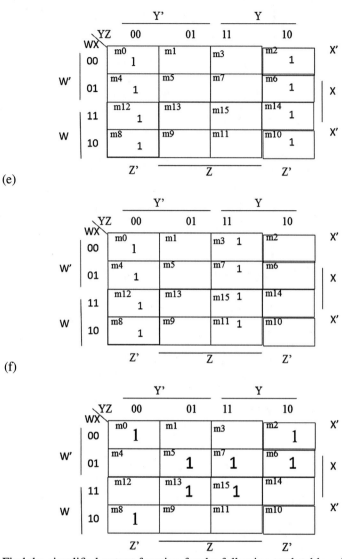

(e)

(f)

4. Find the simplified output function for the following truth table using a K-map:

(a)
X  Y  F
0  0  1
0  1  1
1  0  1
1  1  0

(b)
X  Y  Z  F
0  0  0  1
0  0  1  1

```
0  1  0  0
0  1  1  1
1  0  0  0
1  0  1  1
1  1  0  1
1  1  1  0
```
(c)

| A B C D | F |
|---------|---|
| 0 0 0 0 | 1 |
| 0 0 0 1 | 0 |
| 0 0 1 0 | 1 |
| 0 0 1 1 | 1 |
| 0 1 0 0 | 0 |
| 0 1 0 1 | 1 |
| 0 1 1 0 | 1 |
| 0 1 1 1 | 1 |
| 1 0 0 0 | 1 |
| 1 0 0 1 | 1 |
| 1 0 1 0 | 0 |
| 1 0 1 1 | 0 |
| 1 1 0 0 | 0 |
| 1 1 0 1 | 1 |
| 1 1 1 0 | 1 |
| 1 1 1 1 | 1 |

5. Simplify the following functions using a K-map:

(a) $F(X,Y) = m_2 + m_3$
(b) $F(X,Y) = X + X'Y$
(c) $F(X,Y) = X' + XY'$
(d) $F(X,Y,Z) = m_0 + m_2 + m_5 + m_7$
(e) $F(X,Y,Z) = X'Y'Z' + X'YZ + XY'Z + XYZ$
(f) $F(X,Y,Z) = \pi(0, 2, 5, 7)$
(g) $F(X,Y,Z) = XY'Z + X' + Z + Y'Z'$
(h) $F(W,X,Y,Z) = X'Y'Z' + XYZ' + WXY + W'X'Y' + WZ$
(i) $F(W,X,Y,Z) = X' + XZ' + WX'Y + W'Y' + WZ$

6. Simplify the following functions where D is a *don't care* function:

(a) $F(X,Y,Z) = \Sigma(0,3,4)$
   $D(X,Y,Z) = \Sigma(2,6)$
(b) $F(W,X,Y,Z) = \Sigma(0,1,3,5,9,11)$
   $D(W,X,Y,Z) = \Sigma(2,4,8,10)$

7. Simplify the following functions in the form of SOP and POS, and draw a logic circuit:

(a) $F(X,Y,Z) = \sum(0,2,5,7)$

(b) $F(W,X,Y,Z) = \sum(0,1,4,6,9,11,13,15)$

8. Draw logic circuits for the simplified functions from Problem 6:

(a) Using NAND gates

(b) Using NOR gates

9. Simplify the following function and draw a logic circuit using:

(a) NAND gates

(b) NOR gates

$$F(W,X, Y, Z) = W'X'Z + XY'Z + WX, WY, WX'Y'Z'$$

10. Find the complement of the following functions:

(a) $F(X, Y, Z) = (X' + Y)(X + Z)(Y + Z')$

(b) $F(X, Y, Z) = X'Y + XY'Z + XYZ'$

# Chapter 4
# Combinational Logic

**Objectives: After Completing this Chapter, you Should Be Able to**
- Find output function of a given digital circuit.
- Design a combinational logic circuit using problem description.
- Learn the operation of decoder and its application.
- Learn application of encoder.
- Design and learn the function of a multiplexer.
- Develop half adder, full adder from logic gates.
- Use full adder to design binary adder and subtractor.
- Learn how to design ALU (arithmetic logic unit).
- Use BCD to seven-segment decoder to display a number in decimal.

## 4.1 Introduction

Digital circuit is classified as combinational or sequential logic. Combinational logic is a digital circuit with digital input or inputs and digital output or outputs, this digital circuit performs a specific function, the output of combinational logic depends on current value of inputs, and it is a memoryless circuit, but sequential logic contains memory element. Figure 4.1 shows the block diagram of a combinational logic with inputs and outputs; in combinational logic, the outputs are a function of the inputs.

Consider $F(X,Y) = XY' + X'Y$ which is given in Fig. 4.2; the logic circuit is made of NOT, AND, and OR gates; the output of combinational logic will change by changing the inputs. Table 4.1 shows truth table of Fig. 4.2.

© The Author(s), under exclusive license to Springer Nature Switzerland AG 2022
A. Elahi, *Computer Systems*, https://doi.org/10.1007/978-3-030-93449-1_4

**Fig. 4.1**  Block diagram of a
combinational logic

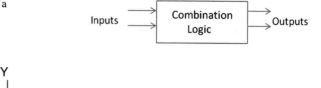

**Fig. 4.2**  Combinational logic circuit

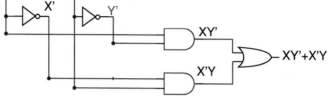

**Table 4.1**  Truth table of
Fig. 4.2

| X | Y | XY' | X'Y | F |
|---|---|-----|-----|---|
| 0 | 0 | 0 | 0 | 0 |
| 0 | 1 | 0 | 1 | 1 |
| 1 | 0 | 1 | 0 | 1 |
| 1 | 1 | 0 | 0 | 0 |

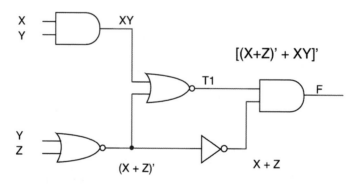

**Fig. 4.3**  Combinational logic

## 4.2   Analysis of Combinational Logic

The objective of the analysis of combinational logic is to find the output function and
truth table of a combinational logic circuit, Fig. 4.3 shows a combinational logic, the
output of the combinational logic is given by function F, and Table 4.2 shows the
truth table of function F.

**Table 4.2** Truth table of
Fig. 4.3

| X Y Z | XY′ | Y′Z | XZ′ | F |
|-------|-----|-----|-----|---|
| 0 0 0 | 0 | 0 | 0 | 0 |
| 0 0 1 | 0 | 1 | 0 | 1 |
| 0 1 0 | 0 | 0 | 0 | 0 |
| 0 1 1 | 0 | 0 | 0 | 0 |
| 1 0 0 | 1 | 0 | 1 | 1 |
| 1 0 1 | 1 | 1 | 0 | 1 |
| 1 1 0 | 0 | 0 | 1 | 1 |
| 1 1 1 | 0 | 0 | 0 | 0 |

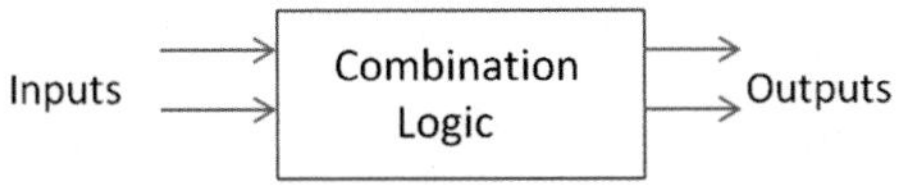

**Fig. 4.4** Block diagram of a
combinational logic

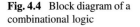

Inputs → Combination Logic → Outputs

$$T1 = \left[(X+Z)' + XY\right]' = (XY)'(X+Z)(XY)' = (X+Z)(X'+Y')$$
$$= XY' + X'Z + Y'Z$$
$$F(X,Y,Z) = (X+Z)T1 = (X+Z)(XY' + X'Z + Y'Z)$$
$$= XY' + XY'Z + XY'Z + X'Z + Y'Z$$
$$F(X,Y,Z) = XY' + XY'Z + X'Z + Y'Z = XY' + Y'Z(X+1) + X'Z$$
$$= XY' + Y'Z + X'Z$$

## 4.3   Design of Combinational Logic

Figure 4.4 shows a block diagram of combinational logic. The following steps show
how to design a combinational logic:

1. Statement of the problem which describes the function of the combinational logic.
2. Define the number of inputs and outputs or maybe they are given by the statement
   of the problem.
3. Assign variables to the inputs and outputs.
4. Develop a truth table by writing all combinations for inputs. The outputs will be
   determined by the statement of the problem.
5. Write the output functions using K-map.
6. Draw a logic circuit.

**Example 4.1**  Design a combinational logic circuit with three inputs and one output;
the output is one when the binary value of the inputs is greater than or equal to three;
otherwise the output will be zero.

### 4.3.1  Solution

Figure 4.4 shows the block diagram of combinational logic with three inputs and one output: the variables X, Y, and Z are assigned to the inputs, and variable F is assigned to the output. Table 4.3 shows truth table for the problem (Fig. 4.5).

In Table 4.3, all input combinations for X, Y, and Z are listed; then according to the statement of the problem, the output F is one when the input is three or more; otherwise the output is zero. The output function F can be represented by the sum of minterms:

$$F(X, Y, Z) = m_3 + m_4 + m_5 + m_6 + m_7 (\text{sum of minterms})$$

By transferring the minterms of F(X,Y,Z) into the K-map as shown in Fig. 4.6 and reading the simplified function from the K-map results F (X,Y,Z) = X + YZ. Figure 4.7 shows the logic circuit for function F (Fig. 4.7).

Table 4.3  The truth table for example 1

| X Y Z | F |
|-------|---|
| 0 0 0 | 0 |
| 0 0 1 | 0 |
| 0 1 0 | 0 |
| 0 1 1 | 1 |
| 1 0 0 | 1 |
| 1 0 1 | 1 |
| 1 1 0 | 1 |
| 1 1 1 | 1 |

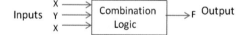

Fig. 4.5  Block diagram of combinational logic of example 4

Fig. 4.6  The K-map for example 4

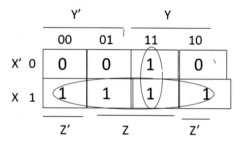

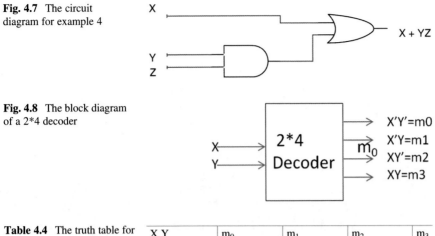

**Fig. 4.7** The circuit diagram for example 4

**Fig. 4.8** The block diagram of a 2*4 decoder

**Table 4.4** The truth table for a 2*4 decoder

| X Y | $m_0$ | $m_1$ | $m_2$ | $m_3$ |
|-----|-------|-------|-------|-------|
| 0 0 | 1 | 0 | 0 | 0 |
| 0 1 | 0 | 1 | 0 | 0 |
| 1 0 | 0 | 0 | 1 | 0 |
| 1 1 | 0 | 0 | 0 | 1 |

## 4.4   Decoder

A decoder is an MSI logic which generates the minterms of a set of inputs; the two variables X and Y generate four minterms, and they are $X'Y' = m_0$, $X'Y = m_1$, $XY' = m_2$, and $XY = m_3$. Figure 4.8 shows the block diagram of a 2*4 decoder (2 inputs and 4 outputs), and Table 4.4 shows the truth table for a 2*4 decoder.

From the truth table of the decoder, the following functions are the outputs of a decoder:

$$m_0 = X'Y', m_1 = X'Y, m_2 = XY', \text{ and } m_3 = XY$$

Figure 4.9 shows logic circuit of 2*4 decoder. Most MSI ICs have an extra input called enable/disable (E/D); the function of the E/D input is to enable or disable an IC as shown in Fig. 4.9. When E/D = 0, all outputs of the decoder will be zeros (meaning the decoder is disabled); the decoder only generates minterms when E/D is set to one.

### 4.4.1   Implementing a Function Using a Decoder

Decoders can be used to design combinational circuits. Consider $F(X,Y) = XY + X'Y'$ which can be implemented using a decoder. The function can be represented by F(X,

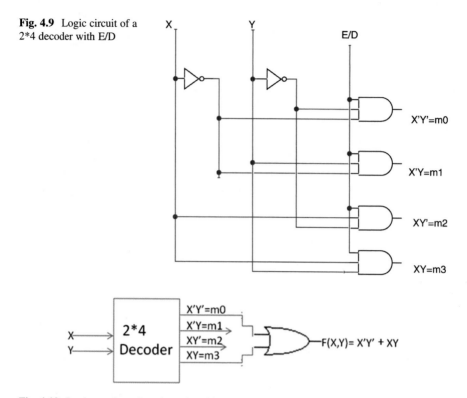

**Fig. 4.9** Logic circuit of a 2*4 decoder with E/D

**Fig. 4.10** Implementing a function using decoder

Y) = $m_3$ + $m_0$. The function contains two variables which are the inputs to the decoder; therefore, a 2*4 decoder is needed, and the output F is the sum of minterms $m_3$ and $m_0$ which is shown in Fig. 4.10.

## 4.5  Encoder

An encoder is the opposite of a decoder; it has $2^n$ inputs and n output, for n = 2 means encoder has $2^2 = 4$ inputs and 2 outputs; the output is binary value of selected input. Figure 4.11 shows block diagram of 4*2 encoder, and Table 4.5 shows the truth table for a 4*2 encoder. In Table 4.5, if A3 = 1, then the output XY = 11, or if A2 = 1, then the output XY = 10.

Figure 4.12 shows the K-map for functions X and Y; the input combinations that are not listed in the truth table of the 2*4 decoder are *don't care*, designated by "X," in both K-maps. Figure 4.13 shows the logic circuit diagram of the 2*4 decoder.

Fig. 4.11 The block diagram of a 4*2 encoder

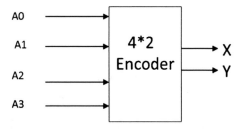

Table 4.5 The truth table for a 4*2 encoder

| A3 | A2 | A1 | A0 | X | Y |
|----|----|----|----|---|---|
| 0 | 0 | 0 | 1 | 0 | 0 |
| 0 | 0 | 1 | 0 | 0 | 1 |
| 0 | 1 | 0 | 0 | 1 | 0 |
| 1 | 0 | 0 | 0 | 1 | 1 |

## 4.6 Multiplexer (MUX)

A MUX is a combinational logic circuit with N inputs and one output; the function of the MUX is to select one of the inputs from many and direct the input to the output. Figure 4.14 shows the basic architecture of a multiplexer. A multiplexer that has N inputs and one output is called an N-to-1 multiplexer. The internal switch selects one input line at a time and transfers that input to the output. When the switch is in position A, it transfers input A to the output; when the switch moves to position B, it transfers input B to the output. This method continues until the switch moves to position D and transfers input D to the output.

The opposite of a multiplexer is a **demultiplexer** (DMUX), shown in Fig. 4.15. The switch moves to send each input to the appropriate output. A DMUX has one input and N outputs— this is called a 1-to-N demultiplexer. When the switch is in position 0, it transfers A to output port 0 and then moves to output port 1 and transfers B to this port. This process continues until the switch moves to output port 3 and transfers D to port 3.

Figure 4.16 shows a 2*1 MUX where A and B are inputs and S is select line; when S = 0, the output of multiplexer is the value of A; when S = 1, the output of MUX is value of B.

Table 4.6 shows truth table of MUX for Fig. 4.16.

$$F\ (S, A, B) = m_3 + m_5 + m_7$$

X

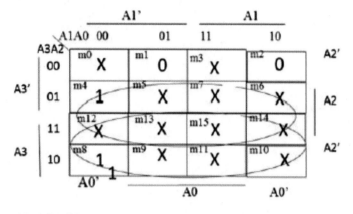

X= A3 +A2

Y

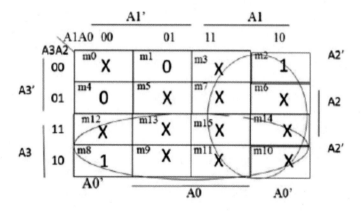

Y =  A1 +A3

**Fig. 4.12** The K-maps for functions X and Y

**Fig. 4.13** The logic circuit diagram of the 4*2 encoder

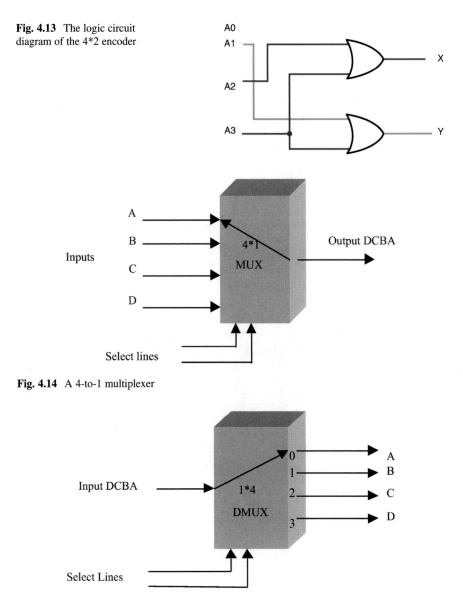

**Fig. 4.14** A 4-to-1 multiplexer

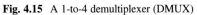

**Fig. 4.15** A 1-to-4 demultiplexer (DMUX)

**Fig. 4.16** Block diagram of
2*1 MX

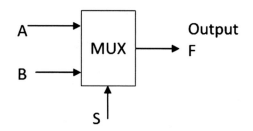

**Table 4.6** Truth table
2*1 MUX

| S | A | B | F |
|---|---|---|---|
| 0 | 0 | 0 | 0 |
| 0 | 0 | 1 | 0 |
| 0 | 1 | 0 | 1 |
| 0 | 1 | 1 | 1 |
| 1 | 0 | 0 | 0 |
| 1 | 0 | 1 | 1 |
| 1 | 1 | 0 | 0 |
| 1 | 1 | 1 | 1 |

**Fig. 4.17** Logic circuit for
2*1 MUX

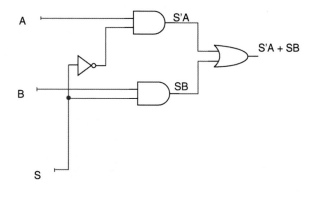

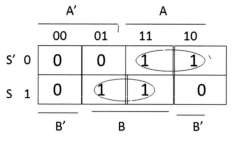

Reading K-map results $F(S,A,B) = S'A + SB$.

Figure 4.17 shows logic circuit for 2*1 MUX.

Figure 4.18 shows a 4*1 MUX where $I_0$, $I_1$, $I_2$, and $I_3$ are the inputs, Y is the output, and $S_0$ and $S_1$ are select lines. Table 4.11 shows the operation of the MUX.

**Fig. 4.18** The block diagram of 4*1 MUX

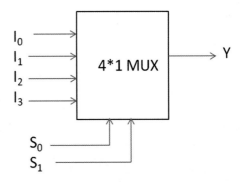

**Table 4.7** The MUX operation

| $S_0$ | $S_1$ | Y |
|-------|-------|-----|
| 0 | 0 | $I_0$ |
| 0 | 1 | $I_1$ |
| 1 | 0 | $I_2$ |
| 1 | 1 | $I_3$ |

The function Y can be generated from Table 4.7; when the inputs $S_0S_1 = 00$, then $Y = I_0$; when $S_0S_1 = 01$, then $Y = I_1$; when $S_0S_1 = 10$, then $Y = I_2$; when $S_0S_1 = 11$, then $Y = I_3$; therefore, the Y output is

$$Y = S_0'S_1'I_0 + S_0'S_1I_1 + S_0S_1'I_2 + S_0S_1I_0$$

Figure 4.19 shows the block diagram for a 4*1 MUX with E/D. Figure 4.20 shows logic circuit of 4*1 MUX; as shown in this figure, the E/D input is added to the logic diagram of MUX; when E/D = 0, the output Y = 0, and the MUX is disabled.

## 4.6.1 Designing Large Multiplexer Using Smaller Multiplexers

A large MUX can be generated by combining small MUX's, an 8*1 MUX can be constructed by two 4*1 multiplexers and one OR gate as shown in Fig. 4.21. In this figure, A, B, and C are select lines. When A = 0, the top MUX is enabled, and when A = 1, the lower MUX is enabled. Also, the 8*1 MUX can be implemented using two 4*1 and one 2*1 MUX as shown in Fig. 4.22. In Fig. 4.22, when A = 0, the output F = Y0, and when A = 1, the output F = Y1.

**Fig. 4.19** Shows block diagram of 4*1 MUX with E/D

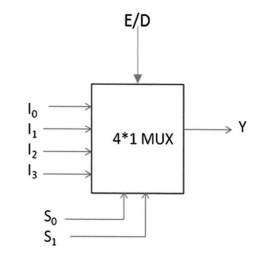

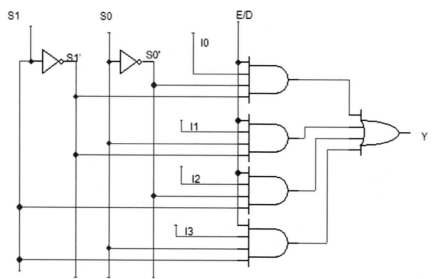

**Fig. 4.20** The circuit logic diagram for a 4*1 MUX

## 4.6.2 Implementing Functions Using Multiplexer

(a) Implementing a three-variable truth table using a 8*1 MUX.

  MUX can be used to implement a digital function; consider truth Table 4.8; the truth table is made of three variables, and it can use an 8*1 MUX to implant the truth table (Table 4.8).

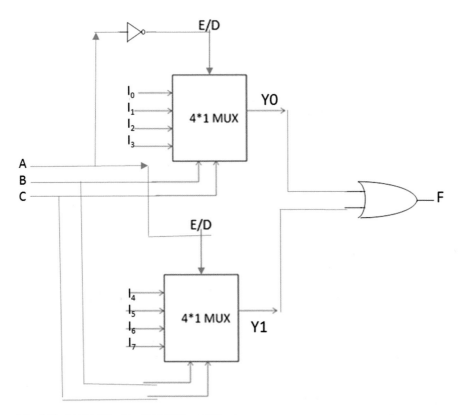

**Fig. 4.21** 8*1 MUX using 4*1 MUX and OR gate

The truth table is made of three variables; therefore, an 8*1 MUX is needed; the A, B, and C variables are connected to the select lines of MUX, and the inputs of the MUX correspond to the output of truth table (Fig. 4.23).

(b) Implementing Table 4.8 using a 4*1 MUX

Number of variables − 1 = number of select lines.

Three variables − 1 = 2 number of select lines for MUX.

If A and B are connected to select lines, when AB = 00, there are two rows in Table 4.9 with AB = 00, then the output F depends on the value of C; in this case C = 0. Consider rows with AB = 01, F = 0 for C = 0, and F = 1 for C = 1; therefore, F = C. Figure 4.24 shows implementation of Table 4.9 using 4*1 MUX.

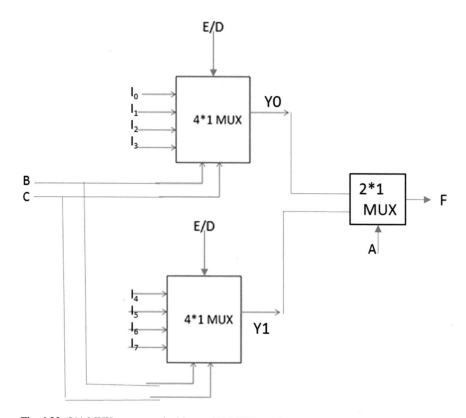

**Fig. 4.22** 8*1 MUX constructed with two 4*1 MUX and 2*1 MUX

**Table 4.8** The truth table for a MUX with three variables

| A | B | C | F |
|---|---|---|---|
| 0 | 0 | 0 | 0 |
| 0 | 0 | 1 | 0 |
| 0 | 1 | 0 | 0 |
| 0 | 1 | 1 | 1 |
| 1 | 0 | 0 | 1 |
| 1 | 0 | 1 | 0 |
| 1 | 1 | 0 | 1 |
| 1 | 1 | 1 | 1 |

## 4.7   Half Adder, Full Adder, Binary Adder, and Subtractor

Half adder (HA) is a logic circuit that adds the bits X and Y; Fig. 4.25 shows block diagram of a HA; the inputs to the HA are X and Y; the outputs of the HA are S (sum) and C (carry). Table 4.10 shows the truth table for a HA. In this truth table, when $X = Y = 1$, then X plus $Y = 10$ results $S = 0$ and $C = 1$.

The functions S and C are

**Fig. 4.23** The block diagram of an 8*1 MUX

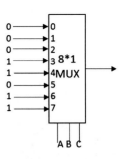

**Table 4.9** Truth table for Fig. 4.24

| A | B | C | F | |
|---|---|---|---|---|
| 0 | 0 | 0 | 0 | 0 |
| 0 | 0 | 1 | 0 | |
| 0 | 1 | 0 | 0 | C |
| 0 | 1 | 1 | 1 | |
| 1 | 0 | 0 | 1 | C′ |
| 1 | 0 | 1 | 0 | |
| 1 | 1 | 0 | 1 | 1 |
| 1 | 1 | 1 | 1 | |

**Fig. 4.24** Multiplexer for Table 4.9

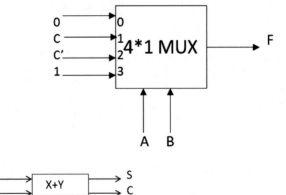

**Fig. 4.25** The block diagram of a half adder

**Table 4.10** The truth table for a half adder

| X Y | C | S |
|---|---|---|
| 0 0 | 0 | 0 |
| 0 1 | 0 | 1 |
| 1 0 | 0 | 1 |
| 1 1 | 1 | 0 |

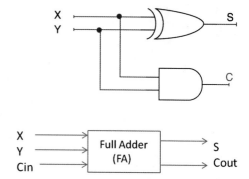

**Fig. 4.26** Logic circuit for HA

**Fig. 4.27** The block diagram of a half adder

**Table 4.11** The truth table for a half adder

| X Y Cin | Cout | S |
|---------|------|---|
| 0 0 0   | 0    | 0 |
| 0 0 1   | 0    | 1 |
| 0 1 0   | 0    | 1 |
| 0 1 1   | 1    | 0 |
| 1 0 0   | 0    | 1 |
| 1 0 1   | 1    | 0 |
| 1 1 0   | 1    | 0 |
| 1 1 1   | 1    | 1 |

$$S = m_1 + m_2 = X'Y + XY' = X \text{ XOR } Y$$

$$C = XY = X \text{ AND } Y$$

Figure 4.26 shows logic circuit of half adder (HA).

### 4.7.1   Full Adder (FA)

Figure 4.27 shows the block diagram of a full adder (FA). The FA will add X + Y + Cin (each is only 1 bit), the outputs of the FA are called S and Cout, and Table 4.11 shows the truth table of a FA; in this table, X, Y, and Cin are added, and the results generate the sum (S) and carry (Cout).

The function of S can be represented by the sum of minterms as shown in Eq. 4.1:

$$S(X, Y, Cin) = X'Y'Cin + X'YCin' + X'YCin + XY'Cin' + XYcin \qquad (4.1)$$

OR

$$S(X, Y, Cin) = Cin \ (X'Y' + XY) + Cin' \ (X'Y + X'Y) \qquad (4.2)$$

$$X'Y + X'Y = X \ XOR \ Y = A$$

$$X'Y' + XY = (X'Y + X'Y)' = A'$$

Therefore, Eq. 4.2 can be written in the form of Eq. 4.3:

$$S(X, Y, Cin) = CinA' + Cin'A = Cin \ XOR \ A \qquad (4.3)$$

Cout can be written in the form of the sum of minterms:

$$Cout \ (X, Y, Cin) = X'YCin, \ + XY'Cin + XYCin' + XYCin$$

OR

$$Cout \ (X, Y, Cin) = Cin \ (X'Y, +XY') + XY \ (Cin' + Cin)$$

OR

$$Cout \ (X, Y, Cin) = Cin \ (X'Y, +XY') + XY$$

Figure 4.28 shows the logic circuit for a full adder.

Figure 4.29 shows block diagram of a full adder (FA), where X, Y, and Cin are the inputs and S and Cout are the outputs.

The full adder can be designed by using two half adders as shown in Fig. 4.30

## 4.7.2 4-Bit Binary Adder

The function of a 4-bit binary adder is to add two 4-bit numbers such as:

$X_3X_2X_1X_0 + Y_3Y_2Y_1Y_0$

When adding $X_0$ by $Y_0$, it results in a sum (S0) and a carry (C0); the C0 is then added to $X_1$ and Y1 which results in a S1 and a C1. Figure 4.31 shows 4-bit binary adder; the Cin is connected to ground to represent zero.

The manufacturer diagram of the 4-bit binary adder is shown in Fig. 4.32 as one IC, and the IC number is 7483.

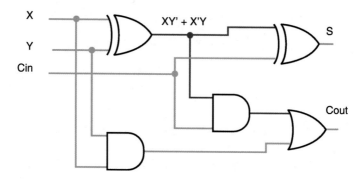

**Fig. 4.28** Full adder logic circuit

**Fig. 4.29** Block diagram of
FA

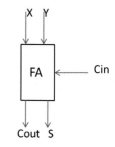

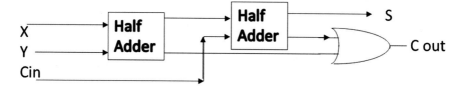

**Fig 4.30** Block Diagram of FA using HA

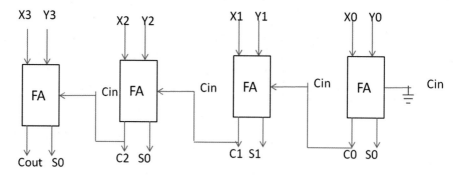

**Fig. 4.31** 4-bit binary adder

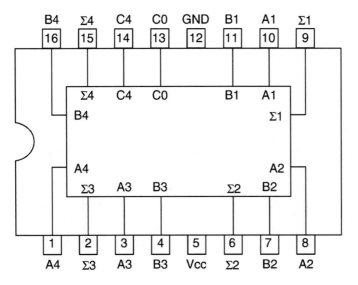

**Fig. 4.32**  7483 4-bit binary adder

### 4.7.3   Subtractor

A subtractor performs the subtraction of $A - B$ or $A + B' + 1$. Figure 4.33 shows a diagram of a subtractor using a 4-bit binary adder. The CI is set to one, and inputs B0, B1, B2, and B3 are complemented.

Figure 4.34 is a modification of Fig. 4.33 which can perform both addition and subtraction. CI is the carry in and CO is the carry out of the 4-bit binary adder. By setting the Add/Sub switch to zero, it performs addition, and by setting Add/Sub switch to one, it performs subtraction.

## 4.8   ALU (Arithmetic Logic Unit)

The function of the **arithmetic logic unit (ALU)** is to perform arithmetic operations such as addition and subtraction and bit-wise logic operations such as AND, OR, and NOT. Figure 4.35 shows the block diagram of an ALU.

In Fig. 4.35, A and B buses are the inputs, and the C bus is the output of the ALU; S1 and S0 are select lines that select the function of the ALU; Table 4.12 shows the function of an ALU; assume A and B are 4 bits and represented by A3, A2, A1, and A0 and B3, B2, B1, and B0.

The select lines of ALU define the size of the multiplexer, since there are 2 select lines; therefore, the size of the MUX is 4*1 (4 inputs and 1 output); the number of bits define the number of multiplexer. A and B are 4 bits; therefore, four 4*1 multiplexers are needed. Figure 4.36 shows a diagram of an ALU.

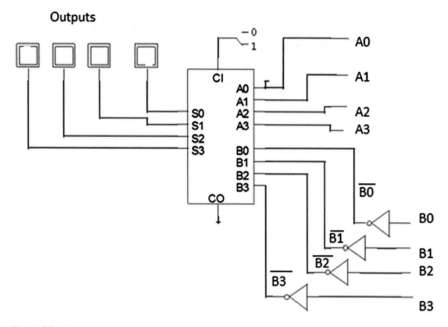

**Fig. 4.33** 4 bit subtractor

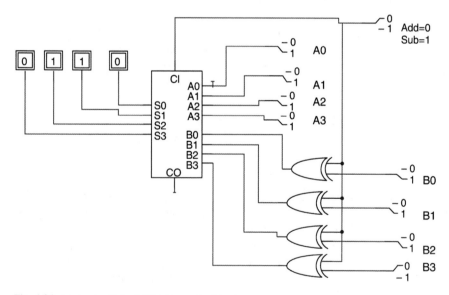

**Fig. 4.34** Logic circuit for 4-bit adder and subtractor

**Fig. 4.35**  Block diagram of
an ALU

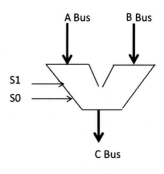

**Table 4.12**  The functions of
an ALU

| S1 | S0 | ALU |
|----|----|-----|
| 0 | 0 | A OR B |
| 0 | 1 | A + B |
| 1 | 0 | A AND B |
| 1 | 1 | A' |

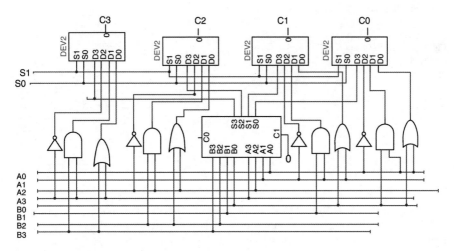

**Fig. 4.36**  A logic circuit diagram of an ALU

## 4.9   Seven-Segment Display

A seven-segment display is made of seven LEDs (light-emitting diode) as shown in
Fig. 4.37; a seven-segment display can display any one digit from zero to nine;
Fig. 4.38 shows the segments that should be on in order to display the digits from
0 through 9.

For displaying 0, all segments should be on except for g; for displaying 8, all
segments should be on. It requires a special decoder called a BCD to seven-segment
decoder to convert binary-coded decimal (BCD) to seven-segment display.

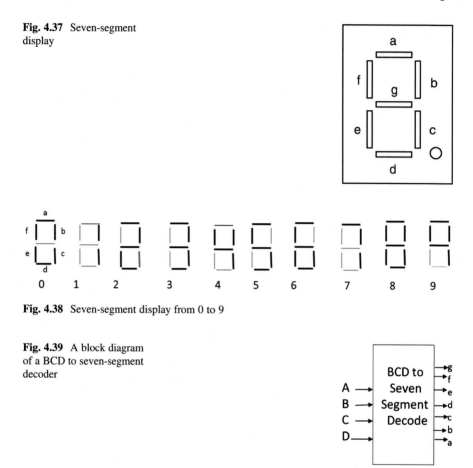

**Fig. 4.37** Seven-segment display

**Fig. 4.38** Seven-segment display from 0 to 9

**Fig. 4.39** A block diagram of a BCD to seven-segment decoder

Figure 4.39 shows a block diagram of the BCD to seven-segment decoder, and Table 4.13 shows the truth table of the BCD to seven-segment decoder. The decoder has 4 inputs and 7 outputs; the input to the decoder is BCD (binary-coded decimal which is from 0000 through 1001), as shown in Table 4.13; if the input value is greater than 1001, then the output is *don't care*.

There are seven outputs, and each output requires a K-map in order to find the output functions. Figure 4.40 shows the K-map for output a.

By reading the K-map of Fig. 4.40, the result is function a:

$$a = B'D' + A + C + BD$$

Using the above procedure, one can find other output functions.

**Table 4.13** The truth table for a BCD to seven-segment decoder

| A | B | C | D | a | b | c | d | e | f | g |
|---|---|---|---|---|---|---|---|---|---|---|
| 0 | 0 | 0 | 0 | 1 | 1 | 1 | 1 | 1 | 1 | 0 |
| 0 | 0 | 0 | 1 | 0 | 1 | 1 | 0 | 1 | 0 | 0 |
| 0 | 0 | 1 | 0 | 1 | 1 | 0 | 1 | 1 | 0 | 1 |
| 0 | 0 | 1 | 1 | 1 | 1 | 1 | 1 | 0 | 0 | 1 |
| 0 | 1 | 0 | 0 | 0 | 1 | 1 | 0 | 0 | 1 | 1 |
| 0 | 1 | 0 | 1 | 1 | 0 | 1 | 1 | 0 | 1 | 1 |
| 0 | 1 | 1 | 0 | 1 | 0 | 1 | 1 | 1 | 1 | 1 |
| 0 | 1 | 1 | 1 | 1 | 1 | 1 | 0 | 0 | 0 | 0 |
| 1 | 0 | 0 | 0 | 1 | 1 | 1 | 1 | 1 | 1 | 1 |
| 1 | 0 | 0 | 1 | 1 | 1 | 1 | 0 | 0 | 1 | 1 |
| 1 | 0 | 1 | 0 | d | d | d | d | d | d | d |
| 1 | 0 | 1 | 1 | d | d | d | d | d | d | d |
| 1 | 1 | 0 | 0 | d | d | d | d | d | d | d |
| 1 | 1 | 0 | 1 | d | d | d | d | d | d | d |
| 1 | 1 | 1 | 0 | d | d | d | d | d | d | d |
| 1 | 1 | 1 | 1 | d | d | d | d | d | d | d |

**Fig. 4.40** The K-map for output a

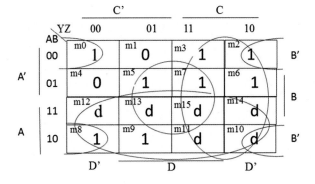

## 4.10 Summary

- Combinational circuit is a digital circuit with one or more digital inputs and digital outputs.
- The outputs of combinational circuit depend on current values of inputs.
- A combinational logic circuit can be represented by a function or a truth table.
- Decoder is a combinational logic with n inputs and $2^n$ outputs; if $n = 2$, then decoder is 2*4.
- Decoder generates minterms of inputs; a decoder with two inputs generates four minterms.
- Multiplexer is a combinational circuit with $2^n$ inputs and one output, where n is number of select lines. If $n = 2$, then multiplexer will be 4*1.
- The function of half adder (HA) is to add 2 bits and generates sum and carry.
- The function of full adder (FA) is to add 3 bits.

- The function of a 4-bit binary adder is to add two 4-bit numbers.
- Arithmetic logic unit (ALU) is a combinational logic that performs arithmetic operations and logic operations.
- Seven-segment displays are used for displaying one digit of decimal number.
- BCD to seven-segment decoder converts 4-bit BCD to 7 bits for interfacing to 7-segment displays.
- Chapter 5 covers sequential logic circuit, and the topics presented in this chapter are D, J-K, and T flip-flops, register, shift register, state diagram, state table, and designing counter.

## Problems

1. The following combinational circuit logic diagrams are given; find the output function and truth table for each function.

   (a)

   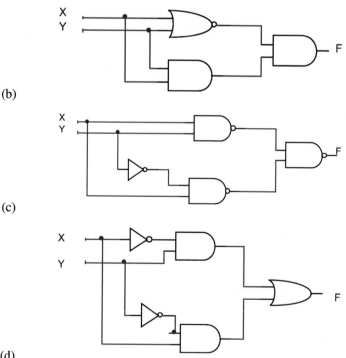

   (b)

   (c)

   (d)

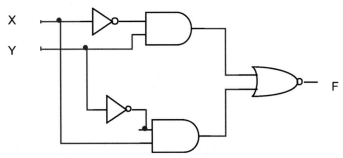

2. Find the output of following gates.

3. Find the output F for each set of inputs.

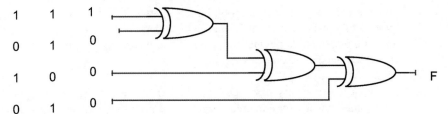

4. Design a logic circuit with three inputs and one output; the output generates even parity bit of the inputs.

    (a) Show the truth table.
    (b) Find output function.
    (c) Draw logic circuit.

5. Implement function $F(X,Y,Z) = XY' + XZ'$ using:

    (a) Decoder
    (b) Multiplexer

6. Implement the following functions using only one decoder and external gates:

    $F1(X,Y,Z) = \sum(0,3,4)$
    $F2(X,Y,Z) = \sum(2,3,5)$

7. Implement a full adder using decoder.
8. The following multiplexer is given; complete its table.

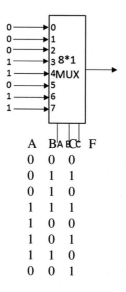

| A | B | C | F |
|---|---|---|---|
| 0 | 0 | 0 |
| 0 | 1 | 1 |
| 0 | 1 | 0 |
| 1 | 1 | 1 |
| 1 | 0 | 0 |
| 1 | 0 | 1 |
| 1 | 1 | 0 |
| 0 | 0 | 1 |

9. Implement function $F(W,X,Y,Z) = \sum(0,1,3,4,7,8,9,11,12,15)$ using MUX.
10. Design an 8-bit binary adder using 4-bit binary adders.
11. Design a 16-bit binary adder using 4-bit binary adders.
12. Design a combinational logic with three inputs and three outputs; if input 0, 1, 2, or 3, then output 3 more than input, if input 4, 5, 6, or 7, then output 3 less than the input.
13. A train with 7 wagons carry passenger and numbered from 1 to 7; each wagon contains a binary switch for emergency, and when any of the switches become on, then display wagon number in conductor cabin in decimal. Verify your design using logisim.
14. Design a combinational circuit with four inputs and one output; the input to the combinational circuit is BCD, and output generates even parity for the input.
15. Design a 16*1 MUX using 4 *1 MUXs.
16. Design a 4-bit ALU to perform the following functions:

    A + B, A − B, A + 1, A′, B′, A OR B, A XOR B, A AND B

17. Design a combinational logic that compares X and Y, where $X = X1X0$ and $Y = Y1Y0$; the output of combinational logic is 1, when $X < Y$; otherwise the output is 0.

    (a) Show truth table.
    (b) Find output function using K-map.

18. Find the output F for the following combinational logic:

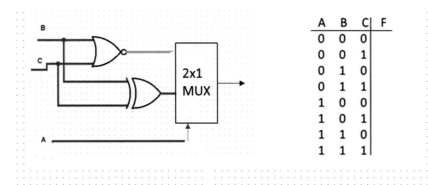

| A | B | C | F |
|---|---|---|---|
| 0 | 0 | 0 | |
| 0 | 0 | 1 | |
| 0 | 1 | 0 | |
| 0 | 1 | 1 | |
| 1 | 0 | 0 | |
| 1 | 0 | 1 | |
| 1 | 1 | 0 | |
| 1 | 1 | 1 | |

19. Design a combinational circuit with three inputs and one output, where the output is one when the binary input has more ones than zeros.

# Chapter 5
# Synchronous Sequential Logic

**Objectives: After Completing this Chapter, you Should Be Able to**
- Analyze the sequential logic.
- Learn operation of the S-R latch.
- Design D flip-flop from S-R latch.
- Learn the application of the D flip-flop.
- Learn operation of the J-K and T flip-flops.
- Design register and shift register using the D flip-flops.
- Develop state table for a sequential circuit.
- Develop state diagram from the state table.
- Develop excitation table for each types of flip-flop.
- Design digital counter.

## 5.1   Introduction

Sequential logic circuit contains memory elements, and the output depends on the current value of input and prior input-level conditions. Figure 5.1shows block diagram of sequential logic as shown in this figure; the outputs depend on the inputs and current state of memory elements; in this figure, the outputs of combinational logic are the inputs to memory elements, and the outputs of memory elements are the inputs to combinational logic; the basic elements of memory elements are flip-flops that can hold binary values as long as the device is powered. The output of a synchronous sequential logic depends on the outputs of memory elements and inputs. The applications of synchronous sequential logic are designing register, counter, and memory. Synchronous sequential logic operates with the clock pulse.

© The Author(s), under exclusive license to Springer Nature Switzerland AG 2022      103
A. Elahi, *Computer Systems*, https://doi.org/10.1007/978-3-030-93449-1_5

**Fig. 5.1** Block diagram of
synchronous sequential
logic

**Fig. 5.2** Block diagram of
S-R latch

**Fig. 5.3** Logic circuit of
S-R latch using NOR gates

**Fig. 5.4** Logic circuit of
S-R latch using NAND
gates

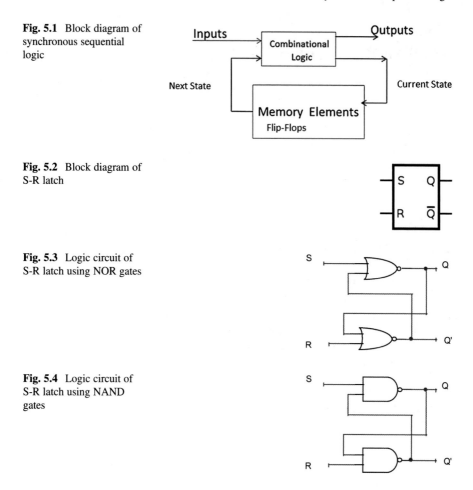

## 5.2  S-R Latch

S-R latch is a type of memory with two inputs S (set) and R (reset), two outputs Q
and Q′, and the outputs are the complement of each other. Figure 5.2 shows block
diagram of a S-R latch.

A S-R latch can be constructed with NOR or NAND gates: Fig. 5.3 shows S-R
latch using NOR gates and Fig. 5.4 shows S-R latch using NAND gates.

### 5.2.1  S-R Latch Operation

Consider S-R latch of Fig. 5.4 which is constructed with NAND gates; the following
steps describe the operation of S-R latch, and Table 5.1 shows its characteristic table:

**Table 5.1** Characteristic table of S-R latch

| S | R | Q | Q' |
|---|---|---|---|
| 0 | 0 | 1 | 1 Forbidden |
| 0 | 1 | 1 | 0 |
| 1 | 1 | 1 | 0 No change |
| 1 | 0 | 0 | 1 |
| 1 | 1 | 0 | 1 No change |

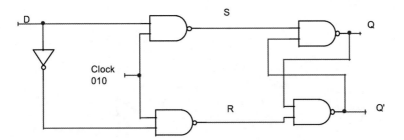

**Fig. 5.5** Logic circuit of D flip-flop

**Table 5.2** Characteristic table of D flip-flop

| Clock | D | Q |
|---|---|---|
| ↑ | 0 | 0 |
| ↑ | 1 | 1 |

1. By setting $S = 0$ and $R = 0$, results outputs $Q = Q' = 1$ which are not permitted because Q and Q' must be a complement of each other; therefore, $S = R = 0$ is prohibited.
2. By setting $S = 0$ and $R = 1$, results $Q = 1$ and $Q' = 0$ as shown in Table 5.1, if S changes from 0 to 1, then Q does not change.
3. By setting $S = 1$ and $R = 0$, results $Q = 0$ and $Q' = 1$, if R changes from 0 to 1, then the Q does not change. It can conclude when $S = R = 1$ the output of Q does not change (if $Q = 0$ stays 0 or $Q = 1$ stays 1). The S-R latch is a basic logic circuit for D, J-K, and T flip-flops.

## 5.3  D Flip-Flop

D flip-flop is a 1-bit memory, and it is used for designing SRAM (Static RAM) and register; Fig. 5.5 shows the logic diagram of D flip-flop. The inputs to flip-flop are D and clock. When the clock is 0, then $S = R = 1$, and according to Table 5.2 the output of flip-flop does not change, by setting D to 0, and changing clock from 0 to 1 results $S = 0$ and $R = 1$ then $Q = 0$ and $Q' = 1$; this means when input $D = 0$ and applying clock, the output Q changes to 0. Setting D to 1 and changing clock from 0 to 1 results $S = 0$ and $R = 1$, and then according to Table 5.2, the Q sets to one; this

**Fig. 5.6** Block diagram of
D flip-flop

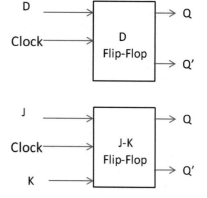

**Fig. 5.7** J-K flip-flop

**Table 5.3** Characteristic
table of J-K flip-flop

| Clock | J | K | Q |
|-------|---|---|---|
| ↑ | 0 | 0 | No change |
| ↑ | 0 | 1 | 0 |
| ↑ | 1 | 0 | 1 |
| ↑ | 1 | 1 | Complement |

means when input D = 1 and applying clock, the output Q changes to 1. Figure 5.6
shows block diagram of D flip-flop, and rising edge of the clock is represented by "''"
and Table 5.2 shows a characteristic table of D flip-flop.

## 5.4  J-K Flip-Flop

Figure 5.7 shows block diagram of a J-K flip-flop where J, K, and clock are the
inputs to the J-K flip-flop. The application of J-K flip-flop is counter and frequency
divider. Table 5.3 shows characteristic table of J-K flip-flop, and the following steps
describe J-K flip-flop operations:

(a) By setting J = K = 0 and applying clock pulse to the flip-flop, the output Q does
    not change, if Q = 0, then stays 0, or if Q = 1, then stays 1.
(b) By setting J = 0, K = 1 and applying clock pulse to the flip-flop, then output Q
    changes to 0.
(c) By setting J = 1, K = 0 and applying clock to the flip-flop, the output Q changes
    to 1.
(d) By setting J = K = 1 and applying a clock pulse, the output of the flip-flop is the
    complement of present output; this means if Q = 0 and applying clock, then
    output changes to 1 and if Q = 1 and applying clock pulse then the output will
    change to 0.

## 5.5   T Flip-Flop

T flip-flop is a special case of J-K flip-flop, and by connecting J and K inputs of J-K flip-flop together results in a T flip-flop; Fig. 5.8 shows a block diagram of T flip-flop, and Table 5.4 shows the characteristic table of T flip-flop; as shown in Table 5.4 if T = 0 and applying clock pulse, then the output of T flip-flop does not change, and if T = 1 and applying clock, then output of flip-flop becomes the complement of the present output.

## 5.6   Register

D flip-flop is 1-bit memory or 1-bit register. If a group of D flip-flops share a common clock, it is called register; the N-bit register is constructed with N D flip-flops, and if 32 D flip-flops use a common clock, then it is called a 32-bit register. Figure 5.9 shows a 4-bit register, and in this figure by placing 1101 at the inputs and applying clock pulse, then the output will be 1101.

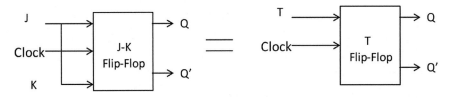

**Fig. 5.8**   Block diagram of T flip-flop

**Table 5.4**   Characteristic table of T flip-flop

| Clock | T | Q |
|-------|---|---|
| ↑ | 0 | No change |
| ↑ | 1 | Complement |

**Fig. 5.9**   4-bit register

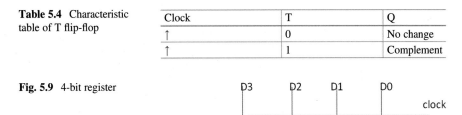

### 5.6.1  Shift Register

A shift register has one serial input, and one bit is loaded from serial input into the register by every clock pulse, and then each bit of register is shifted to the next bit position. Figure 5.10 shows 4-bit shift right register operation, and after shifting one bit to the right, the contents of the register will be 0101. Figure 5.11 shows 4-bit serial shift right register.

**Fig. 5.10**  4-bit shift right register operation

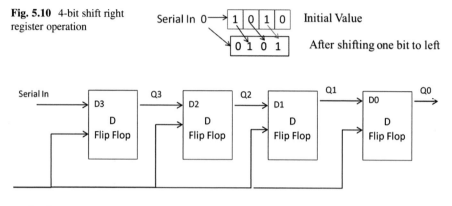

**Fig. 5.11**  4-bit serial shift right register

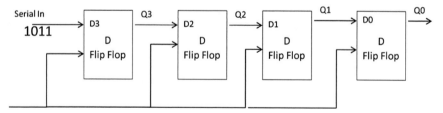

| Clock# | Q3 | Q2 | Q1 | Q0 |               |
|--------|----|----|----|----|---------------|
|        | 0  | 0  | 0  | 0  | Initial values |
| 1      | 1  | 0  | 0  | 0  |               |
| 2      | 1  | 1  | 0  | 0  |               |
| 3      | 0  | 1  | 1  | 0  |               |
| 4      | 1  | 0  | 1  | 1  |               |

**Fig. 5.12**  4-bit shift right register

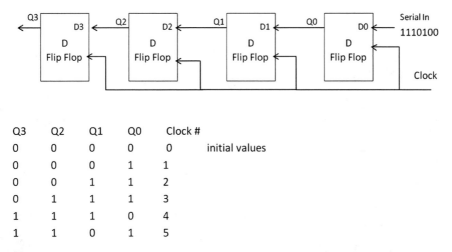

| Q3 | Q2 | Q1 | Q0 | Clock # | |
|----|----|----|----|---------|--|
| 0 | 0 | 0 | 0 | 0 | initial values |
| 0 | 0 | 0 | 1 | 1 | |
| 0 | 0 | 1 | 1 | 2 | |
| 0 | 1 | 1 | 1 | 3 | |
| 1 | 1 | 1 | 0 | 4 | |
| 1 | 1 | 0 | 1 | 5 | |

**Fig. 5.13** 4-bit shift left register

| **Table 5.5** Barrel shifter operation | S1 | S0 | D3 | D2 | D1 | D0 |
|---|----|----|----|----|----|----|
| | 0 | X | A3 | A2 | A1 | A0 |
| | 1 | 0 | 0 | A3 | A2 | A1 |
| | 1 | 1 | A2 | A1 | A0 | 0 |

**Example** Figure 5.12 shows 4-bit shift right register; shows the contents of register after applying four clock pulses, and assumes that initial output of each D flip-flop is zero.

Figure 5.13 shows shift left register with serial input of 1110100, and Table 5.5 shows contents of register after applying five clock pulses; assumes the initial output of each flip-flop is 0.

### 5.6.2   Barrel Shifter

Barrel shifter is used for shifting data left and right; barrel shifter uses combinational logic rather shift register; combinational logic does not require clock and it is the fastest shifter; Fig. 5.14 shows 4-bit barrel shifter and Table 5.5 shows operation table for barrel shifter.

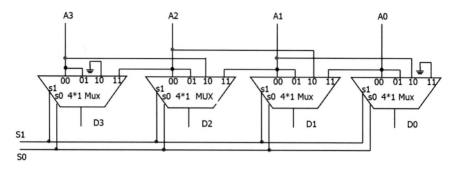

**Fig. 5.14**  4-bit barrel shifter

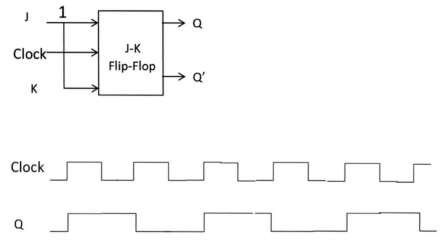

**Fig. 5.15**  Frequency divider using J-K flip-flop

## 5.7   Frequency Divider Using J-K Flip-Flop

Figure 5.15 shows a J-K flip-flop as frequency divider; the inputs J and K are set to 1 and assume the initial value of $Q = 0$; as shown in this figure, for every two clock pulses applied to the flip-flop, then Q generates one clock pulse as shown in Fig. 5.15; this means the circuit divides the frequency by 2.

## 5.8   Analysis of Sequential Logic

In a combinational logic, the truth table represents the characteristic of a function, but characteristic of sequential logic is represented by state table. State table is represented by present state, next state, and output. The present state is the current state of the flip-flop (current output), and the next state is the output of flip-flop after

**Fig. 5.16** Sequential logic

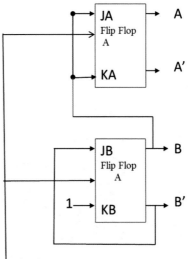

Clock

**Table 5.6** State table for Fig. 5.16

| Present state | | Next state | |
|---|---|---|---|
| A | B | A | B |
| 0 | 0 | 0 | 1 |
| 0 | 1 | 1 | 0 |
| 1 | 0 | 1 | 1 |
| 1 | 1 | 0 | 0 |

applying clock to the sequential logic. Figure 5.16 shows a sequential circuit with two J-K flip-flops with two states A and B; the state table consists of two columns: present state and next state; present state represents current outputs of the flip-flops with all possible values for A and B (00, 01, 10, and 11) as shown in Table 5.6.

Consider the first row; the present state is 00 (means A = 0 and B = 0); it is interested to find the outputs of flip-flops (next state) by applying clock pulse. If A = 0 and B = 0 result in JA = KA = 0, JB = KB = 1, then applying clock pulse to flip-flops results A = 0 and B = 1 (AB = 01 next state).

Consider the second row, for present state A = 0 and B = 1; therefore, JA = KA = 1, JB = 0, and KB = 1; applying a clock pulse to the flip-flops results to next state with A = 1 and B = 0; the same procedure is used for row 10 and 11 to find the next states. Table 5.6 shows state table for sequential logic of Fig. 5.16.

**Example 5.1** Find the state table of Fig. 5.17.

Figure 5.17 contains external input X, and Table 5.7 shows stable with two columns for next state: one for X = 0 and another one for X = 1.

**Fig. 5.17**  Sequential circuit

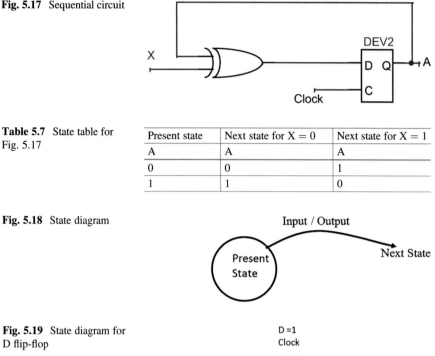

**Table 5.7**  State table for Fig. 5.17

| Present state | Next state for X = 0 | Next state for X = 1 |
|---|---|---|
| A | A | A |
| 0 | 0 | 1 |
| 1 | 1 | 0 |

**Fig. 5.18**  State diagram

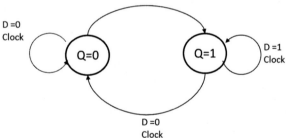

**Fig. 5.19**  State diagram for D flip-flop

## 5.9   State Diagram

Another way to represent the characteristic of sequential logic is by state diagram, as shown in Fig. 5.18. The present state is the current value of flip-flop, and applying clock the present state changes to next state.

### 5.9.1   D Flip-Flop State Diagram

Figure 5.19shows state diagram of D flip-flop, and the following steps describe state diagram:

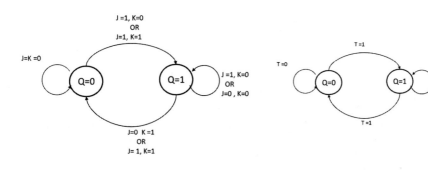

J-K Flip Flop State Diagram                             T-Filip flop State
diagram

**Fig. 5.20**  J-K and T flip-flops state diagrams

**Table 5.8** Excitation table for D flip-flop

| Q(t) | Q(t + 1) | D |
|------|----------|---|
| 0 | 0 | 0 |
| 0 | 1 | 1 |
| 1 | 0 | 0 |
| 1 | 1 | 1 |

1. If Q = 0 (present state), by setting D = 1 and applying clock, then Q changes from 0 to 1 (next state).
2. If Q = 1, by setting D = 1 and applying clock, then output stays 1.
3. If Q = 1, by setting D = 0 and applying clock, then output changes to 0.
4. If Q = 0, by setting D = 0 and applying clock, then output stays 0.

Figure 5.20 shows state diagram of J-K and T flip-flops.

## 5.10  Flip-Flop Excitation Table

The application of excitation table is to determine input or inputs of a flip-flop in order to get predefined output.

### 5.10.1  D Flip-Flop Excitation Table

Table 5.8 shows excitation table for D flip-flop: $Q(t)$ is present output (present state) and $Q(t + 1)$ is the next state.

## 5.10.2  Excitation Table Operation

Consider the first row of Table 5.8, if $Q(t) = 0$ (present state) and it is desired after applying a clock pulse, the output $Q(t + 1)$ stays 0 and then D must be set to 0.

Consider the second row, the current output of D flip-flop (present state) is 0, and it is desired to change the output $Q(t + 1)$ to one; therefore, the input D must set to 1.

Consider the third row, the present output is 1, and it is desired to change the output (next state) to 0; therefore the input D must be set to 0.

Consider the fourth row, the present state is 1, and it is desired to stay 1; therefore, D must set to one.

## 5.10.3  J-K Flip-Flop Excitation Table

Table 5.9 shows J-K flip-flop excitation table, and the following steps describe how this table was generated:

1. Consider the first row of excitation table, the present state of the flip-flop is zero, and it desired to stay 0 by applying clock pulse; therefore, J must set to zero and K is don't care (0 or 1).
2. Consider the second row, the present state of the flip-flop is 0, and it is desired to change the output to 1 by applying clock pulse; therefore, J must set to 1 and K can don't care.
3. Consider the third row, the present state $Q(t)$ is 1, and it is desired to change it to 0; therefore, the J can don't care and $k = 1$.
4. Consider the fourth row, the present state is 1, and it is desired to stay 1; therefore, J can don't care and $K = 0$.

## 5.10.4  T Flip-Flop Excitation Table

Table 5.10 shows excitation table of T flip-flop.

**Table 5.9** J-K flip-flop excitation table

| $Q(t)$ | $Q(t + 1)$ | J | K |
| --- | --- | --- | --- |
| 0 | 0 | 0 | d |
| 0 | 1 | 1 | d |
| 1 | 0 | d | 1 |
| 1 | 1 | d | 0 |

**Table 5.10** T flip-flop excitation table

| Q(t) | Q(t + 1) | T |
|------|----------|---|
| 0 | 0 | 0 |
| 0 | 1 | 1 |
| 1 | 0 | 1 |
| 1 | 1 | 0 |

**Fig. 5.22** Sequential logic
of 2-bit counter

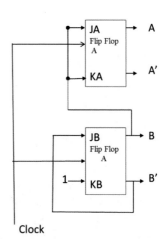

## 5.11  Counter

A counter is a sequential logic which is used to count the number of pulses applied
to it or divide a clock frequency if a system has a clock of 16 Hz, and it is possible to
use a counter to change 16 Hz clock to 4 Hz. The following steps describe how to
design a counter:

(a) Define count sequence which is a sequence that the counter will count.
(b) Use count sequence to determine the number of flip-flops.
(c) Select the types of flip-flop.
(d) Use count sequence to develop state table.
(e) Use state table and flip-flop excitation table to develop excitation table for
    counter.
(f) Use K-map to find the input functions or function to each flip-flop.
(g) Draw the sequential logic for the counter.

Example: Design a counter to count 0—1—2—3 and repeat using J-K flip-flops.
The biggest number in count sequence is 3 which is represented in binary by 11;
therefore, two flip-flops are needed and it is called A and B as shown in Fig. 5.22,
and Table 5.11 shows state table for the counter.
The present state defines the current output of flip-flops, and the next state is the
output of flip-flops after applying a clock pulse.

**Table 5.11** State table of counter

| Present state | | Next state | |
|---|---|---|---|
| A | B | A | B |
| 0 | 0 | 0 | 1 |
| 0 | 1 | 1 | 0 |
| 1 | 0 | 1 | 1 |
| 1 | 1 | 0 | 0 |

**Table 5.12** Excitation table for counter

| Present state | | Next state | | JA | KA | JB | KB |
|---|---|---|---|---|---|---|---|
| A | B | A | B | | | | |
| 0 | 0 | 0 | 1 | 0 | d | 1 | d |
| 0 | 1 | 1 | 0 | 1 | d | d | 1 |
| 1 | 0 | 1 | 1 | d | 0 | 1 | d |
| 1 | 1 | 0 | 0 | d | 1 | d | 1 |

Table 5.12 shows excitation table for the counter which was developed by using excitation table of JK flip-flop.

Consider the first row, the present output of J-K flip-flops is 00 (A = 0, B = 0), and it is desired the outputs change to 01 (A = 0 and B = 1); therefore, it must set JA = 0, KA = d (don't care) in order for the A to stay 0 and set JB = 1, KB = d in order for B to change from 0 to 1.

Consider the second row, the present state is 01 (A = 0 and B = 1), and it is desired the output changes to 10 (A = 1 and B = 0); therefore, it must set JA = 1, KA = d, and JB = 0, KB = d.

It is desired to find the input functions to the flip-flops, the present state are the inputs, and JA, KA, JB, and KB are the outputs of the Table 5.12, by transferring the outputs to the K-maps, and reading the K-maps results the input functions to the flip-flops; Fig. 5.21 shows K-maps for JA, KA, JB, and KB.

The input functions to the flip-flops are JA = JB = B, JB = B' and KB = 1, and Fig. 5.22 shows the circuit of 2-bit counter.

## 5.12  Summary

- Sequential logic circuit requires clock to operate.
- A S-R latch is the basic component for flip-flop.
- S-R latch can be constructed by the NAND or NOR gates.
- The basic element of sequential logic is flip-flop.
- Flip-flop is a memory element with the two outputs Q and Q'.

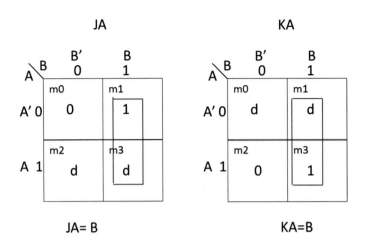

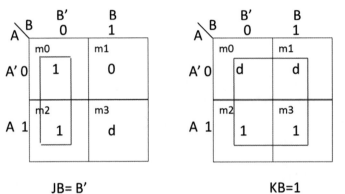

**Fig. 5.21** K-maps for 2-bit counter

- By applying clock to the D flip-flop, the value of the D input copies at the Q output.
- D flip-flop use for designing register.
- Register is a group of the D flip-flop sharing the same clock.
- J-K flip-flop is used for the designing counter.
- Connecting inputs of a J-K flip-flop together results a T flip-flop.
- State table and state diagram show operation of a sequential circuit.
- Chapter 6 is an introduction to computer architecture which covers basic components of a microcomputer as well as CPU technologies, CPU architectures, multicore processor, instruction execution steps, pipelining, and microcomputer buses.

## Problems

1. Complete the following table for D flip-flop.

| D | Q(t) present output | Q(t + 1) next output |
|---|---|---|
| 0 | 0 | |
| 0 | 1 | |
| 1 | 0 | |
| 1 | 1 | |

2. Complete the following table for J-K flip-flop.

| J | K | Q(t) present output | Q(t + 1) next output |
|---|---|---|---|
| 0 | 0 | 0 | |
| 0 | 0 | 1 | |
| 0 | 1 | 0 | |
| 0 | 1 | 1 | |
| 1 | 0 | 0 | |
| 1 | 0 | 1 | |
| 1 | 1 | 0 | |
| 1 | 1 | 1 | |

3. Complete the following table for T flip-flop.

| T | Q(t) present output | Q(t + 1) next output |
|---|---|---|
| 0 | 0 | |
| 0 | 1 | |
| 1 | 0 | |
| 1 | 1 | |

4. The following figure shows a sequential logic; complete the following table assuming initial value of Q1 = 0 and Q2 = 0. Use logisim to verify your answer.

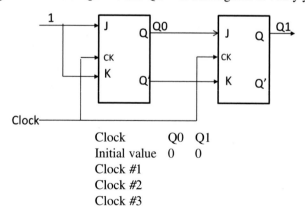

| Clock | Q0 | Q1 |
|---|---|---|
| Initial value | 0 | 0 |
| Clock #1 | | |
| Clock #2 | | |
| Clock #3 | | |

5. Show an 8-bit register using D flip-flops.

6. The following shift register given, find the output after five clock pulses.

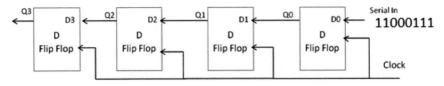

7. With the following sequential logic given, assume initial value for Q0 = 0 and Q1 = 0, and flip-flop changes state in rising edge of clock pulse; complete the following table and then use logisim to verify your result.

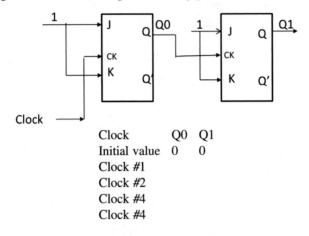

| Clock | Q0 | Q1 |
|---|---|---|
| Initial value | 0 | 0 |
| Clock #1 | | |
| Clock #2 | | |
| Clock #4 | | |
| Clock #4 | | |

8. Complete the following excitation table for J-K flip-flop.

| Q(t) | Q(t + 1) | J | K |
|---|---|---|---|
| 0 | 0 | | |
| 0 | 1 | | |
| 1 | 0 | | |
| 1 | 1 | | |

9. Design a counter to count 0—1—2—3—4—5—6—7 and repeat.

   (a) Use J-K flip-flops.
   (b) Use T flip-flops.
   (c) Verify your design using logisim.

10. Find the state diagram for the following state table.

| | AB | AB |
|---|---|---|
| AB | X = 0 | X = 1 |
| 00 | 01 | 10 |
| 01 | 10 | 00 |
| 10 | 11 | 01 |
| 11 | 00 | 10 |

11. What is the content of the following register after shifting five times to the left?

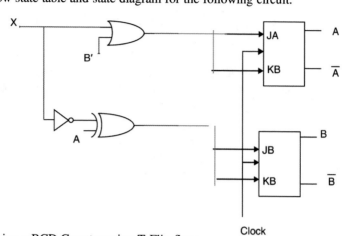

12. Show state table and state diagram for the following circuit.

13. Design a BCD Counter using T-Flip flops.

# Chapter 6
# Introduction to Computer Architecture

**Objectives: After Completing this Chapter, you Should be Able to:**
- List the components of a microcomputer.
- List the components of CPU.
- Distinguish CPU technologies.
- Learn architecture of multicore processor.
- Compare the RISC processor with the CISC processor.
- Explain the difference between the von Neumann and the Harvard architecture.
- Distinguish between the 32-bit processor and the 64-bit processor.
- Explain the instruction execution steps.
- Show advantage of the instruction pipelining.
- Distinguish different types of the microcomputer buses.
- Explain operation of the USB bus.

## 6.1 Introduction

Just as the architecture of a building defines its overall design and functions, so computer architecture defines the design and functionality of a computer system. The components of a microcomputer are designed to interact with one another, and this interaction plays an important role in the overall system operation.

### 6.1.1 Abstract Representation of Computer Architecture

A general view of the layered architecture of computer is shown in Fig. 6.1. In this figure, it shows that the Instruction Set architecture is located between hardware and software as it acts as the interface layer.

© The Author(s), under exclusive license to Springer Nature Switzerland AG 2022
A. Elahi, *Computer Systems*, https://doi.org/10.1007/978-3-030-93449-1_6

**Fig. 6.1** Layered
architecture of a computer

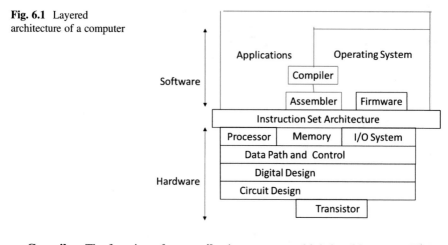

**Compiler:** The function of a **compiler** is to convert a high-level language (HLL) to assembly language, and the assembler then converts assembly language to machine code (binary). Some of the most popular HLLs are C, C++, Java, and Python.

**Firmware:** Firmware is a software program or set of instructions programmed on a hardware device. It provides the necessary instructions for how the device communicates with the other computer hardware devices such as the BIOS, device manager, and device driver.

**Instruction Set Architecture**: The instruction set are the set of different commands that are supported by the hardware. It is basically the interface between the hardware and software layers.

**Applications:** Applications are programs run by users like text editors and web browsers.

The hardware layer consists of the CPU, memory, and I/O devices that make up the computer.

**Operating System:** Operating systems manage computer hardware resources such as input/output operations, managing memory, and scheduling processes for execution.

## 6.2   Components of a Microcomputer

A standard microcomputer consists of a microprocessor (CPU), buses, memory, serial input/output, programmable I/O interrupt, and direct memory access DMA. Figure 6.2 shows components of microcomputer.

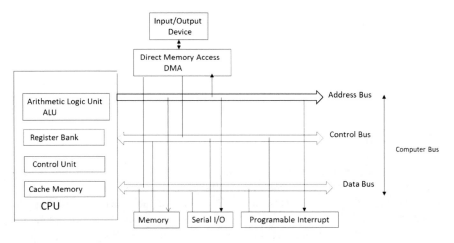

**Fig. 6.2** Components of a microcomputer

## *6.2.1 Central Processing Unit (CPU)*

The *central processing unit* (*CPU*) is the "brain" of the computer and is responsible for accepting data from input devices, processing the data into information, and transferring the information to memory and output devices. The CPU is organized into the following three major sections:

1. Arithmetic logic unit (ALU)
2. Control unit
3. Registers

The function of the *arithmetic logic unit* (*ALU*) is to perform arithmetic operations such as addition, subtraction, division, and multiplication and logic operations such as AND, OR, and NOT.

The function of the *control unit* is to control input/output devices, generate control signals to the other components of the computer such as read and write signals, and perform instruction execution. Information is moved from memory to the registers; the registers then pass the information to the ALU for logic and arithmetic operations.

### 6.2.1.1 Register Bank

Register is the fastest memory in a computer which holds information.

### 6.2.2   CPU Buses

When more than one wire carries the same type of information, it is called a bus. The most common buses inside a microcomputer are the address bus, the data bus, and the control bus.

#### 6.2.2.1   Address Bus

The address bus defines the number of addressable locations in a memory IC by using the $2^n$ formula, where n represents the number of address lines. If the address bus is made up of three lines, then there are $2^3 = 8$ addressable memory locations, as shown in Fig. 6.3. The size of the address bus directly determines the maximum numbers of memory locations that can be accessed by the CPU. For example, a CPU with 32 address bus can have $2^{32}$ addressable memory locations.

#### 6.2.2.2   Data Bus

The data bus is used to carry data to and from the memory. In Fig. 6.3, each location can hold only 8 bits. The size of a memory IC is represented by $\mathbf{2^n \times m}$ where n is the number of address lines and m is the size of each location, usually each memory location holds one byte. In Fig. 6.3, where $n = 3$ and $m = 8$, the size of the memory is

$$2^3 * 8 = 64 \text{ bits.}$$

The size of data bus plays an important factor on CPU performance, current CPU's data bus is 32 bits or 64 bits, and a CPU with 32-bit data bus means it can read or write 32 bits of data in and from memory.

Early generation of CPU contains 8-bit data bus, and each memory location holds one byte, for reading word "book" as shown in Fig. 6.4. The CPU requires to access memory four times.

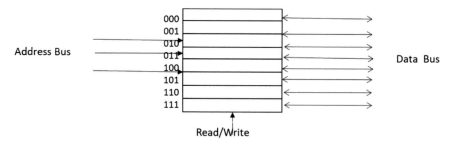

**Fig. 6.3**  A memory with three address lines and four data lines

**Fig. 6.4**  CPU with 8 bit
Data Bus

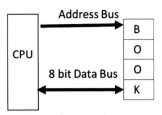

By increasing data bus from 8 bits to 32 bits, then CPU can access memory and read entire word "book." Most CPUs offer instruction to read 1 byte, 2 bytes, or 4 bytes from memory. Most CPUs can read from memory or write in memory in sizes of one byte, two bytes, and four bytes at a time.

### 6.2.2.3   Control Bus

The control bus carries control signals from the control unit to the computer components in order to control the operation of each component. In addition, the control unit receives control signals from computer components. Some of the control signals are as follows:

| | |
|---|---|
| *Read signal*: | The read line is set to high to read from memory location or input/output (I/O) devices |
| *Write signal*: | The write line is used to write data into the memory |
| *Interrupt*: | Indicates an interrupt request |
| *Bus request*: | The device is requesting to use the computer bus |
| *Bus Grant*: | Gives permission to the requesting device to use the computer bus |
| *I/O Read and Write*: | I/O read and write are used to read from or write to I/O devices |

## 6.2.3   Memory

There are four types of memory used in a computer: registers, cache memory, main memory, and secondary memory.

**Register: Registers are located inside CPU and hold.**

**Cache memory**: Cache memory is part of the CPU and it is the fastest type of memory. Cache is made of SRAM (Static Random Access Memory).

**Main Memory**: Main memory is a type of DRAM (Dynamic Random Access Memory) and is volatile and fast.

**Secondary Memory**: Secondary memory is also called disk or secondary storage, and includes mass storage devices like hard disks or solid-state drives (SSD). This type of memory is slow and less expensive when compared with DRAM.

### 6.2.4   Serial Input/Output

**To interface with other devices, computers need to use different kinds of serial input and output interfaces.** The most popular devices that utilize input/output interfaces include USB devices and PCI Express.

### 6.2.5   Direct Memory Access (DMA)

*Direct memory access (DMA)* allows for the transfer of blocks of data from memory to an I/O device or vice versa. Without DMA, the CPU reads data from memory and writes it to an I/O device. Transferring blocks of data from memory to an I/O device requires the CPU to do one read and one write for each operation. This method of data transfer takes a lot of time. The function of DMA is to transfer data from memory to an I/O device directly, without using the CPU, so that the CPU is free to perform other functions.

The DMA performs the following functions in order to use the computer bus:

- The DMA sends a request signal to the CPU.
- The CPU responds to the DMA with a grant request, permitting the DMA to use the bus.
- The DMA controls the bus and the I/O device is able to read or write directly to or from memory.
- The DMA is able to load a file off an external disk into main memory when large blocks of data need to be transferred to a sequential range of memory. DMA is much faster and more efficient than a CPU.

### 6.2.6   Programmable I/O Interrupt

When multiple I/O devices such as external drives, hard disks, printers, monitors, and modems are connected to a computer as shown in Fig.6.5, a mechanism is necessary to synchronize all device requests. The function of a programmable interrupt is to check the status of each device and inform the CPU of the status of each; for example, the printer is not ready, a disk is write protected, this is an unformatted disk, and there is a missing connection to a modem. Each device

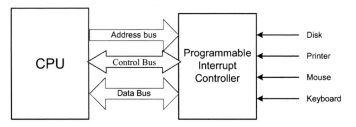

**Fig. 6.5**  Programmable interrupt controller

sends a signal to the programmable I/O interrupt controller in order to update its status. Figure 6.5 shows the programmable I/O interrupt controller.

### 6.2.7  32-Bit Versus 64-Bit CPU

The size of register plays an important role in the performance of CPU. A 32-bit processor can perform operations on 32-bit data; therefore, the size of registers is 32 bits and ALU also performs 32-bit operations. A 64-bit CPU performs operation in 64-bit data; therefore, it contains 64-bit register and 64-bit ALU.

Most desktop and server computers are using AMD and Intel processors; they might use 32 bits or 64 bits. Intel and AMD processor use the same architecture; this means a program in computer with Intel processor can run on a computer with AMD processor.

## 6.3  CPU Technology

There are two types of technology used for designing CPU and they are called CISC and RIS.

### 6.3.1  CISC (Complex Instruction Set Computer)

In 1978, Intel developed the 8086 microprocessor chip. The 8086 was designed to process a 16-bit data word; it had no instruction for floating point operations. At the present time, the Pentium processes 32-bit and 64-bit words, and it can process floating point instructions. Intel designed the Pentium processor in such a way that it can execute programs written for earlier 80 × 86 processors.

The characteristics of 80 × 86 are called complex instruction set computers (CISC), which include instructions for earlier Intel processors. Another CISC processor is VAX 11/780, which can execute programs for the PDP-11 computer. The CISC processor contains many instructions with different addressing modes, for example, the VAX 11/780 has more than 300 instructions with 16 different address modes.

The major characteristics of CISC processor are as follows:

1. A large number of instructions.
2. Many addressing modes.
3. Variable length of instructions.
4. Most instruction can manipulate operands in the memory.
5. Control unit is microprogrammed.

### 6.3.2   RISC

Until the mid-1990s, computer manufacturers were designing complex CPUs with large sets of instructions. At that time, a number of computer manufacturers decided to design CPUs capable of executing only a very limited set of instructions.

One advantage of reduced instruction set computer is that they can execute their instructions very fast because the instructions are simple. In addition, the RISC chip requires fewer transistors than the CISC chip. Some of the RISC processors are the PowerPC, MIPS processor, IBM RISC System/6000, ARM, and SPARC.

The major characteristics of RISC processors are as follows:

1. Require few instructions.
2. All instructions are the same length (they can be easily decoded).
3. Most instructions are executed in one machine clock cycle.
4. Control unit is hardwired.
5. Few address modes.
6. A large number of registers.

RISC processor uses hardware and CISC processor microprogram for control unit, and the control unit with hardware uses less space in a CPU; therefore, the designer of CPU can add more registers to RISC processor compared with CISC.

The advantage of CISC processor is that designer can add new instruction without changing the architecture of the processor. Table 6.1 shows the comparison of CISC and RISC.

**Table 6.1** Comparison of RISC and CISC processor

| CISC | RISC |
| --- | --- |
| Variable instruction length | Fixed instruction length |
| Variable opcode length | Fixed opcode length |
| Memory operands | Load/store instructions |
| Example: Pentium | ARM, MIPS |

## 6.4 CPU Architecture

There are two types of CPU architecture and they are *von Neumann and Harvard architecture*.

### 6.4.1 Von Neumann Architecture

It is a program consisting of code (instructions) and data. Figure 6.6 shows a block diagram of the von Neumann architecture. Von Neumann uses the data bus to transfer data and instructions from the memory to the CPU.

### 6.4.2 Harvard Architecture

Harvard architecture uses separate buses for instructions and data as shown in Fig. 6.7. The instruction address bus and instruction bus are used for reading instructions from memory. The address bus and data bus are used for writing and reading data to and from memory.

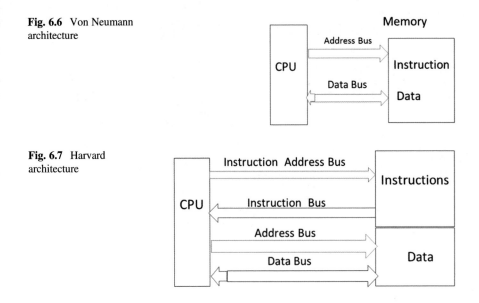

**Fig. 6.6** Von Neumann architecture

**Fig. 6.7** Harvard architecture

## 6.5   Intel Microprocessor Family

Intel designs and manufactures microprocessors for microcomputers. Each processor has a number or name, which is used by the computer designer to access the information provided by the manufacturer of the processor.

Intel microprocessor IC numbers and names are 8088, 80,286, 80,386, 80,486, Pentium, Pentium II, Pentium III, and Pentium IV which they called IA-86 (Intel architecture-86). Recently, Intel and HP developed a new processor called Itanium which is a 64-bit processor. The following is a list of the characteristics of Intel microprocessor (Table 6.2):

Most workstations or laptops use CPU which is manufactured by Intel and AMD Corporations. Intel processor is classified by the IA-16 (Intel architecture 16-bit processor), IA-32, and IA-64.

### 6.5.1   Upward Compatibility

Intel architecture is upward compatible meaning a program written for IA-16 processor can run on IA-32.

Figure 6.8 shows general register IA-16, where AH, AL, BH, BL, CH, CL, DH, and DL are 8-bit registers and AX, BX, CX, and DX are 16-bit register. The AX, BX, CX, and DX are combinations of two registers.

**Table 6.2**  Characteristics of Intel microprocessor

|                  | 80486dx          | Pentium                        | Pentium Pro                     | Pentium Pro II                   | Pentium II                        |
|------------------|------------------|--------------------------------|---------------------------------|----------------------------------|-----------------------------------|
| Register size    | 32 bits          | 32 bits                        | 32 bits                         | 32/64 bits                       | 32/64 bits                        |
| Data bus size    | 32 bits          | 64 bits                        | 64 bits                         | 64 bits                          | 64 bits                           |
| Address size     | 32 bits          | 32 bits                        | 32 bits                         | 32 bits                          | 32 bits                           |
| Max memory       | 4 GB             | 4 GB                           | 4 GB                            | 4 GB                             | 4 GB                              |
| Clock speed      | 25, 33 MHz       | 60, 166 MHz                    | 150, 200 MHz                    | 233, 340, 400 MHz                | 450, 500 MHz                      |
| Math processor   | Built-in         | Built-in                       | Built-in                        | Built-in                         | Built-in                          |
| L1 cache         | 8 KB, 16 KB      | 8 KB instruction 8 KB data     | 8 KB instruction 8 KB data      | 16 KB instruction 16 KB data     | 16 KB instruction 16 KB data      |
| L2 cache         | No               | No                             | 256 KB or 512 KB                | 512 KB                           | 512 KB                            |
| MMX technology   | No               | No                             | Yes                             | Yes                              | Yes                               |

L1 cache is the cache memory built inside the microprocessor.
L2 cache is not part of microprocessor; it is in a separate IC

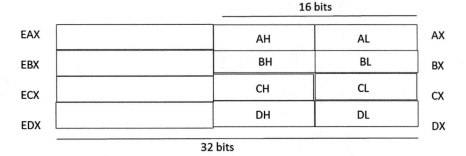

**Fig. 6.8** IA-16 registers

**Fig. 6.9** Intel IA-32-bit registers

Figure 6.9 shows IA-32 general registers where EAX, EBX, ECX, and EDX are general registers, and also Fig. 6.9 also contains IA-16-bit registers; therefore, a program was written for IA-16 can be executed by IA-32.

## 6.6  Multicore Processors

A *multicore processor* is an integrated circuit (IC) with two or more independent CPU which is called core, and they are executing multiple instructions simultaneously in order to increase performance. A quad-core processor is a chip with four independent units called cores that read and execute instructions such as add, move data, and branch. Figure 6.10 shows a block diagram of quad-core processors which are all sharing a memory. The following are some of the multicore processors:

Two cores (dual-core CPUs) such as AMD Phenom II X2 and Intel Core Duo
Three cores (tri-core CPUs) such as AMD Phenom II X3
Four cores (quad-core CPUs) such as AMD Phenom II X4, Intel's i5 and i7 processors
Six cores (hexa-core CPUs) such as AMD Phenom II X6 and Intel Core i7 Extreme Edition 980X

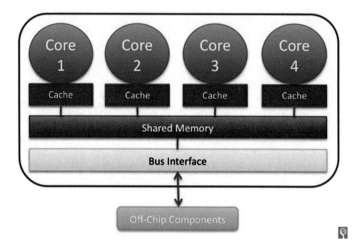

**Fig. 6.10**  Multicore processor architecture

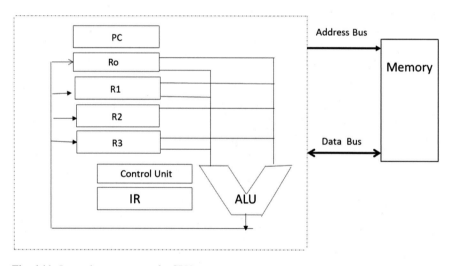

**Fig. 6.11**  Internal components of a CPU

Eight cores (octa-core CPUs) such as Intel Core i7 5960X Extreme Edition and
   AMD FX-8350
Ten cores (deca-core CPUs) such as Intel Xeon E7-2850

## 6.7 CPU Instruction Execution Steps

Figure 6.11 shows the internal components of a CPU, which consists of data registers R0, through R3, PC (Program Counter), IR (Instruction Register), ALU, and Control Unit.

The PC (Program Counter) holds the address of the next instruction to be executed in a program, and the IR (Instruction Register) holds that instruction.

In general, the CPU performs the following steps to execute one instruction:

1. **Fetch instruction (IF):** The Fetch instruction is when the operation system uploads a program into the memory of the computer, then stores the starting address of the program on the PC. Then the control unit places the contents of the PC onto the address bus, reads the instruction from memory, and stores it in the IR. The PC will then increment for the next instruction to be fetched. In summary, the moving of an instruction from memory to the IR is called the Fetch instruction.
2. Decode instruction (D): Now that the IR holds the instruction (such as ADD R3, R2,R1, or AND R3,R1,R2), the control unit will decode the type of instruction stored and move the specified register contents into inputs of the ALU.
3. Execute instruction (E): With ALU registers loaded with the correct data, the instruction will be executed using the specified operation.
4. Write results (R): After execution, the results of the calculation (output of ALU) are stored into a register or memory.

### 6.7.1 Pipelining

Pipelining will increase the performance of CPU meaning executing more instructions in less time.

Figure 6.12 shows execution of four instructions without pipelining; in this figure, CPU executes one instruction at a time, and each stage takes T second; then total execution time is 16 T.

Figure 6.13 shows execution of four instructions using pipeline; CPU fetches I1 instruction and moves it to decode unit; CPU while decoding I1 will fetch I2 instruction and this process will continue.

As shown in Fig. 6.13, it takes 7 T to complete the execution of four instructions. Figure 6.13 shows that at time T4 CPU writes the results of execution I1 into memory and at the same time fetches instruction I4 from memory, but it is

| F1 | D1 | E1 | W1 | F2 | D2 | E2 | W2 | F3 | D3 | E3 | W3 | F4 | D4 | E4 | W4 |

Instruction 1          Instruction 2          Instruction 3          Instruction 4

**Fig. 6.12** Execution of instruction without pipeline

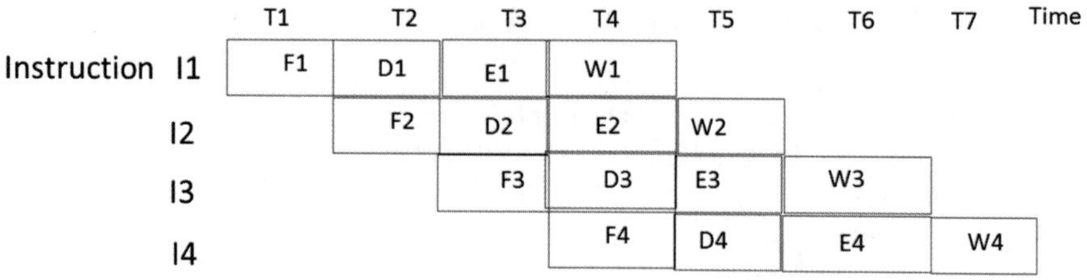

| | T1 | T2 | T3 | T4 | T5 | T6 | T7 | Time |
|---|---|---|---|---|---|---|---|---|
| Instruction I1 | F1 | D1 | E1 | W1 | | | | |
| I2 | | F2 | D2 | E2 | W2 | | | |
| I3 | | | F3 | D3 | E3 | W3 | | |
| I4 | | | | F4 | D4 | E4 | W4 | |

**Fig. 6.13** Execution of instruction using pipeline

impossible to read and write at the same time into or from memory; therefore, having two separate caches (instruction cache and data cache) will overcome this conflict, and this type of architecture is called Harvard architecture.

## 6.8  Disk Controller

The disk controller moves the disk drive head, reads, and/or writes data. The most popular disk controllers are IDE (integrated disk electronics) and **SATA (Serial Advanced Technology Attachment) that** is a computer bus interface that connects host bus adapters to mass storage devices such as hard disk.

## 6.9  Microcomputer Bus

There are currently a number of different computer buses on the market that are designed for microcomputers. Some of the computer BUS are ISA, MCA, EISA, VESA PCI, FireWire, USB, and PCI Express. Universal serial bus (USB) and PC Express are covered in more detail because they are more advanced than other buses.

### 6.9.1  ISA Bus

The *industry standard architecture (ISA) bus* was introduced by IBM for the IBM PC using an 8088 microprocessor. The ISA bus has an 8-bit data bus and 20 address lines at a clock speed of 8 MHz. The PC AT type uses the 80,286 processor which has a 16-bit data bus and 24-bit address lines and is compatible with the PC.

## 6.9.2   Microchannel Architecture Bus

The *microchannel architecture (MCA) bus* was introduced by IBM in 1987 for its PS/2 microcomputer. The MCA bus is a 32-bit bus that can transfer four bytes of data at a time and runs at a 10 MHz clock speed. It also supports 16-bit data transfer and has 32-bit address lines. Microchannel architecture was so expensive, the non-IBM vendors developed a comparable but less expensive solution called the EISA bus.

## 6.9.3   EISA Bus

The *extended ISA (EISA) bus* is a 32-bit bus that also supports 8- and 16-bit data transfer bus architectures. EISA runs at 8-MHz clock speeds and has 32-bit address lines.

## 6.9.4   VESA Bus

The *video electronics standard association (VESA) bus*, which is also called a video local bus (VL-BUS), is a standard interface between the computer and its expansion. As applications became more graphically intensive, the VESA bus was introduced to maximize throughput of video graphics memory. The VESA bus provides fast data flow between stations and can transfer up to 132 Mbps.

## 6.9.5   PCI Bus

The *peripheral component interconnect (PCI) bus* was developed by Intel Corporation. PCI bus technology includes a 32-/64-bit bus that runs at a 33/66 MHz clock speed. PCI offers many advantages for connections to hubs, routers, and network interface cards (NIC). In particular, PCI provides more bandwidth: up to 1 gigabit per second as needed by these hardware components.

The PCI bus was designed to improve the bandwidth and decrease latency in computer systems. Current versions of the PCI bus support data rates of 1056 Mbps and can be upgraded to 4224 Mbps. The PCI bus can support up to 16 slots or devices in the motherboard. Most suppliers of ATM (asynchronous transfer mode) and 100BaseT NICs offer a PCI interface for their products. The PCI bus can be expanded to support a 64-bit data bus. Table 6.3 compares different bus architectures showing characteristics of ISA, EISA, MCA, VESA, and PCI buses. Figure 6.14 shows the PCI bus.

**Table 6.3** Characteristics of various buses

| Bus type | ISA | EISA | MCA | VESA | PCI | PCI-64 |
|---|---|---|---|---|---|---|
| Speed (MHz) | 8 | 8.3 | 10 | 33 | 33 | 64 |
| Data bus bandwidth (bits) | 16 | 32 | 32 | 32 | 32 | 66 |
| Max. data rate (MB/s)[a] | 8 | 32 | 40 | 132 | 132 | 508 |
| Plug and play capable | No | No | Yes | Yes | Yes | Yes |

[a]*MB/s* megabytes/second

**Fig. 6.14** PCI card

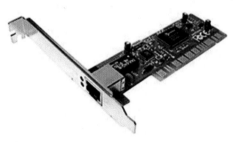

## 6.9.6    *Universal Serial BUS (USB)*

The *universal serial bus (USB)* is a computer serial bus which enables users to connect peripherals such as the mouse, keyboard, modem, CD-ROM, scanner, and printer, to the outside of a computer without any configuration. Personal computers equipped with USB will allow the user to connect peripherals to the computer, and the computer will automatically be configured as the devices are attached to it. This means that a USB has the capability to detect when a device has been added or removed from a PC. USB is a true plug-and-play bus. Up to 127 peripherals can be connected to a PC with a USB. USB version 1.1 was released in 1998 which supports data rate of 12 Mbps (full speed) and 1.5 Mbps (low speed); the low speed is used for devices such mouse, keyboards, and joysticks. The USB version 2 is a high speed (480 Mbps) that is compatible with USB 1.1 The USB 2.0 specification was developed by seven leading computer manufacturers and it was announced in 1999. The maximum cable length for USB is 5 m.

## 6.9.7    *USB Architecture*

Figure 6.15 shows the USB architecture; the USB system is logically a tree topology but physically is a star topology because each USB device communicates directly with the Root Hub. There is only one host controller in any USB system.

A USB system consists of USB host controller, USB root hub, USB hub, USB cable, USB device, client software, and host controller software.

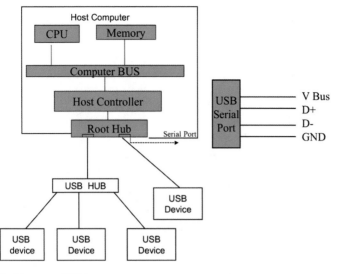

**Fig. 6.15** Architecture of USB

### 6.9.7.1 Host Controller

The host controller initiates all data transfer, and root hub provides a connection between devices and host controller. The root hub receives transaction generated by host controller and transmits to the USB devices. The host controller uses polling to detect a new device and is connected to the bus or disconnected from. Also, USB host controller performs the following functions:

(a) Host controller sets up the device for operation (device configuration).
(b) Packet generation.
(c) Serializer/deserializer.
(d) Process request from device and host.
(e) Manage USB protocol.
(f) Managing flow between host and USB devices.
(g) Assign address to the devices.
(h) Execute client software.
(i) Collecting status bit from USB ports.

### 6.9.7.2 Root Hub

The root hub performs power distribution to the devices, enables and disables the ports, and reports status of each port to the host controller. The root hub provides the connection between the host controller and USB ports.

### 6.9.7.3   Hub

Hubs are used to expand the number of devices connected to the USB system. Hubs are able to detect when a device is attached or removed from port. Figure 6.16 shows the architecture of hub. The upstream port is connected to the host, and USB devices are connected to downstream port.

In downstream transmission, all devices that are connected to the hub will receive the packet, but only the device accepts the packet that the device address matches with address in the token. In upstream transmission, the device sends the packet to the hub, and hub transmits the packet to its upstream port only. USB improves its speed and every few years a new version was developed. Table 6.4 shows different versions of USB.

### 6.9.7.4   USB Cable

Figure 6.15 shows USB port with four pins, which consists of four wires, with the V bus used to power the devices. D+ and D− are used for signal transmission.

### 6.9.7.5   USB Device

USB device is divided into the classes such as hub, printer, or mass storage. The USB device has information about its configuration such as class, type, manufacture ID, and data rate. Host controller uses this information to load device software from the hard disk. USB device might have multiple functions such as a volume in a speaker. Each function in a USB device is specified by the endpoint address.

**Fig. 6.16** Architecture of hub

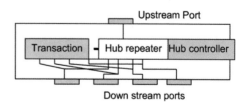

Down stream ports

**Table 6.4** USB version and its data rate

| USB version | Release data | Data rate | Data rate designation |
|---|---|---|---|
| USB 1.0 | 1996 | 1.5 Mbps | Low speed |
| USB 1.1 | 1998 | 12 Mbps | Full speed |
| USB 2.0 | 2000 | 480 Mbps | High speed |
| USB 3.0 | 2008 | 5 Gbps | Super speed |
| USB 3.1 | 2013 | 10 Gbps | Super speed + |

## 6.9.8 PCI Express Bus

PCI express was introduced in mid-1990 with 33 MHz frequency, and during the time the speed of BUS was increased to 66 MHz. Due to new development in networking technology such as Gigabit Ethernet and I/O devices that demand more bandwidth, there is a need for a new bus technology with higher bandwidth. The PCI express was approved by Special Interest Group in 2002, and chipset starts shipping in 2004. The PCI express has the following features:

- PCI express is point-to-point connection between devices.
- PCI express is a serial bus.
- PCI express uses pocket and layer architecture.
- Compatible with PCI bus through software.
- End-to-end link data integrity (error detection).
- Isochronous data transfer.
- Selectable bandwidth.

### 6.9.8.1 PCI Express Architecture

Figure 6.17 shows PCI express architecture, and the function of host bridge is to interface CPU bus with memory and PCI express switch. The switch is used to increase the number of PCI's express ports.

### 6.9.8.2 PCI Express Protocol Architecture

Figure 6.18 shows PCI express protocol architecture. The protocol consists of PCI software, transaction, data link, and physical layer.

**Fig. 6.17** PCI express architecture

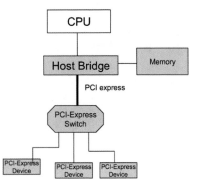

**Fig. 6.18** PCI express
protocol architecture

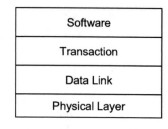

**Fig. 6.19** PCI express
connections

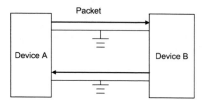

### 6.9.8.3   Software Layer

The software layer is used for compatibility with PCI, initialization, and enumeration of the devices connected to the PCI express.

### 6.9.8.4   PCI Express Physical Layer

Figure 6.19 shows two devices which are connected through PCI express link (lane); each lane is made of four wires, and each PCI express lane consists of two simplex connections, one for transmitting the packet and another one for receiving the packet. The PCI express lane supports 2.5 giga transfer/s each direction.

PCI express link may configure X1, X2, X4, X4, X16, and X32 lane, where X1 means 1 lane with 4 wires and X4 means 4 lanes with 16 wires and finally X32 means 32 lanes with 128 wires. PCI-32 means PCI express with 32 lanes. The clock for PCI express serial link is embedded into the data by using 8B/10B encoding.

## 6.10   FireWire

FireWire or IEEE 1394 is a high-speed serial bus used for connecting digital devices such as a digital video or camcorder. The bus is able to transfer data at the rate of 100, 200, or 400 Mpbs. The IEEE 1394 cable consists of six copper wires; two of the wires carry power and four of the wires carry signal as illustrated in Table 6.5. Some FireWire connectors come with four pins, without having power pins. Figure 6.20 shows FireWire male and female connectors.

**Table 6.5**  IEEE 1394 pins

| Pin | Signal name | Description |
|-----|-------------|-------------|
| 1 | Power | Unregulated DC; 17–24 V no load |
| 2 | Ground | Ground return for power and inner cable shield |
| 3 | TPB− | Twisted-pair B, differential signals |
| 4 | TPB+ | Twisted-pair B, differential signals |
| 5 | TPA− | Twisted-pair A, differential signals |
| 6 | TPA+ | Twisted-pair A, differential signals |

**Fig. 6.20**  FireWire male
and female connectors

Female                          Male

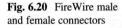

**Fig. 6.21**  HDMI connectors

## 6.10.1    HDMI (High-Definition Multimedia Interface)

HDMI is an interface between two devices for transferring uncompressed video data and compressed or uncompressed digital audio data. Some of the applications of HDMI are computer monitor, digital TV, and video projector. Figure 6.21 shows different types of HDMI connectors.

### 6.10.1.1    Motherboard

The motherboard is a printed circuit board (PCB) that contains most components of a computer such as CPU, RAM, ROM, expansion slots, and USB. Figure 6.22 shows an image of motherboard.

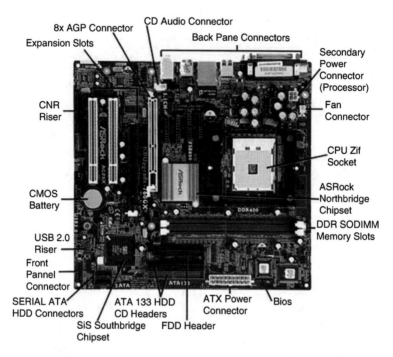

**Fig. 6.22** Image of motherboard

## 6.11  Summary

- The components of a microcomputer are the CPU, memory, parallel I/O, serial I/O, programmable interrupt, and DMA.
- The function of the CPU is executing instruction.
- The components of the CPU are the arithmetic logic unit (ALU), control unit, and registers.
- Most computers use three types of memory: cache memory (SRAM), main memory (DRAM or SDRAM), and secondary memory (hard disk, tape drive, and floppy disk).
- Semiconductor memory types are DRAM, SDRAM, EDORAM, DDR, SDRAM, RDRAM, ROM, and EPROM.
- SRAM is used in the cache memory; DRAM and SDRAM are used in the main memory.
- SATA, SCSI-1, SCSI-2, and SCSI-3 are the computer peripheral controllers.
- ISA bus, EISA, MCA, and PCI are the microcomputer buses, and FireWire is a high-speed serial bus with a data rate up to 400 Mbps.
- USB is a serial bus.
- PCI express is a serial bus.
- Chapter 7 covers semiconductor memory, hard disk, solid-state drive, cache memory mapping methods, and virtual memory.

# Review Questions

• *Multiple choice questions*

1. The function of the _____ is to perform arithmetic operations.

   (a) Bus
   (b) Serial port
   (c) ALU
   (d) Control unit

2. When you compare the functions of a CPU and a microprocessor, _____.

   (a) They are the same
   (b) They are not the same
   (c) The CPU is faster than microprocessor
   (d) The microprocessor is faster than CPU

3. RISC processors use _____.

   (a) Complex instruction sets
   (b) Reduced instruction sets
   (c) (a) and (b)
   (d) None of the above

4. The CISC processor control unit is _____.

   (a) Hardware
   (b) Microcode
   (c) (a) and (b)
   (d) None of the above

5. Direct memory access allows for the transfer of blocks of data from memory to an I/O device (or vice versa) without using the _____.

   (a) CPU
   (b) Data bus
   (c) Control bus
   (d) DMA controller

6. Which of the following buses are 32-bit?

   (a) ISA
   (b) PCI and EISA
   (c) EISA and ISA
   (d) MCA and ISA

7. How many Memory location does have a memory with 12 Address Lines

   (a) 1024
   (b) 2048
   (c) 4096
   (d) 256

8. How many Memory location does have a memory with 16 Address Lines

   (a) 1K
   (b) 4K
   (c) 64K
   (d) 32K

9. How many Memory location does have a memory with 22 Address Lines

   (a) 10K
   (b) 1M
   (c) 2M
   (d) 4M

10. The Fetch instruction means

   (a) Executing Instruction
   (b) Read Instruction from memory
   (c) Decode Instruction
   (d) Store Data

- Short Answer Questions

   1. Show Abstract of a Computer?
   2. What is function of OS?
   3. What is function Compiler?
   4. What is function of Assembler?
   5. What is application of Firmware?
   6. List the components of a microcomputer.
   7. Explain the functions of a CPU.
   8. List the functions of an ALU.
   9. What is the function of a control unit?
   10. List components of a CPU?
   11. How many bits a half word?
   12. How many bits is a word?
   13. Explain the function of an address bus and a data bus.
   14. Explain the function of DMA.
   15. What is the application of a parallel port?
   16. What is the application of a serial port?
   17. What is maximum memory for a CPU with 16 address lines and 8 data lines?
   18. List the types of memory use in a computer.
   19. Whatis the type of memory use for cache memory?

20. What is the type of memory use for main memory?
21. What are the type of memory use for secondary memory?
22. What are the characteristics of a 32- bit machine?
23. What are the characteristics of a 64- bit machine?
24. 24.List characteristics of CISC processor.
25. List characteristics of RISC processor.
26. Distinguish between von Neumann architecture and Harvard architecture.
27. What is the advantage of multicore processor versus single core?
28. List CPU instruction execution steps.
29. Explain fetch instruction
30. Explain the decode instruction.
31. What does the PC (register) stand for, and what is its function?
32. What does IR stand for?
33. How long does it take for a CPU to execute five instructions using pipelining if each stage of pipeline takes 20 min.
34. Calculate execution time for question#33 using non-pipeline processor.
35. List types of disk controller.
36. List two computer buses.
37. List two serial buses.
38. What is maximum number USB ports a computer can have?
39. Show pin connection of USB port.
40. Show diagram of PCIe lane.
41. What is the application of FireWire?
42. What is application of HDMI?

# Chapter 7
# Memory

**Objectives: After Completing this Chapter, you Should Be Able to**
- Distinguish different types of semiconductor memory.
- Explain sector, track on hard disk.
- To calculate disk capacity.
- Learn memory hierarchy.
- Explain cache miss, cache hit, and cache hit ratio.
- Describe types of memory use in a computer.
- Explain different mapping methods use in cache memory.
- Translate virtual address to physical address.
- Explain function of page table.
- Generate physical address from virtual address.

## 7.1   Introduction

In a computer, memory holds instructions (code) and data, memory plays an important part of a computer performance, and register is a type of memory with small capacity. There are two types of memory used in a computer, and they are classified as semiconductor memory and hard disk. Semiconductor memory can be volatile or non-volatile memory. Volatile memory loses its contents when power is removed from it, while non-volatile memory will keep its contents without power.

## 7.2   Memory

Computer memory can be classified as volatile and non-volatile memory.

**Volatile Memory**: It requires power in order to hold information and they are Random Access Memory (RAM) and Static Random-Access Memory

A. Elahi, *Computer Systems*, https://doi.org/10.1007/978-3-030-93449-1_7

(SRAM). Volatile memory is used for temporary storage and it is faster than non-volatile memory.

**Non-Volatile Memory**: Non-volatile memory does not require power to hold information and they are used for long time storage such as Read-Only Memory (ROM), Non-Volatile Random-Access Memory (NVRAM) such as Flash Derive, Hard Disk, and Solid-State Drive (SSD).

## 7.2.1  RAM

Data can be read from or written into *random-access memory (RAM)*. The RAM can hold the data as long as power is supplied to it and it is called *volatile memory*. Figure 7.1 shows a general block diagram of RAM consisting of a data bus, address bus, and read/write signals. The data bus carries data out from or into the RAM. The address bus is used to select a memory location. The read signal becomes active when reading data from RAM, and the write line becomes active when writing to the RAM. Remember, RAM can only hold information when it has power. Figure 7.2 shows a 16 * 8 bit RAM or $2^4$ * 8 bit or 16 B RAM.

In Fig. 7.2, the address is 4 bits; therefore, there are $2^4 = 16$ memory locations, and if each location holds 1 B, then there is 16 B of memory, a memory with m address lines, then there is $2^m$ memory locations. Table 7.1 shows the number of address lines and equivalent decimal number of memory locations.

There are many types of RAM such as *dynamic RAM (DRAM)*, *synchronous DRAM (SDRAM)*, *EDO RAM*, *DDR SDRAM*, *RDRAM*, and *static RAM (SRAM)*.

- *Dynamic RAM (DRAM)* is used in main memory. DRAM uses fewer components to make one bit; therefore, it can design DRAM integrated circuit (IC) with large capacity as 4 GB per IC; Fig. 7.3 shows one bit DRAM.

The cell capacitor can be charged with logic one or zero, but it requires to be refreshed (recharged) about every 1 ms. The CPU cannot read from or write to memory while the DRAM is being refreshed; this makes DRAM the slowest running memory.

**Fig. 7.1** RAM block diagram

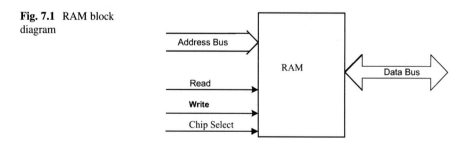

**Fig. 7.2** 16 bytes of RAM

Address

| Address | |
|---------|----------|
| 0000 | 10101011 |
| 0001 | 11001100 |
| 0010 | 10000001 |
| 0011 | 10000000 |
| 0100 | 11111111 |
| 0101 | 11111000 |
| 0110 | 10000000 |
| 0111 | 00000000 |
| 1000 | 00111111 |
| 1001 | 00000001 |
| 1010 | 01010101 |
| 1011 | 10000000 |
| 1100 | 00000011 |
| 1101 | 10000011 |
| 1110 | 11100000 |
| 1111 | 11000000 |

**Table 7.1** Number of address and memory locations

| Number of addresses | Number of memory locations | Representation |
|---------------------|----------------------------|----------------|
| 10 | $2^{10} = 1024$ | 1 K |
| 11 | $2^{11} = 2048$ | 2 K |
| 12 | $2^{12} = 4096$ | 4 K |
| 13 | $2^{13} = 8192$ | 8 K |
| 14 | $2^{14} = 16,384$ | 16 K |
| 16 | $2^{16} = 65,536$ | 64 K |
| 20 | $2^{20} = 1,048,576$ | 1 M |
| 24 | $2^{24} = 16,777,261$ | 16 M |
| 32 | $2^{32} = 4,294,967,296$ | 4 G |

**Fig. 7.3** 1-bit DRAM

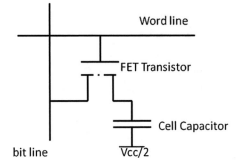

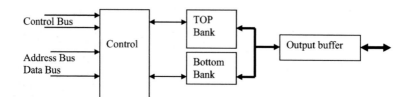

**Fig. 7.4** Block diagram of SDRAM

**Fig. 7.5** Rambus memory module (Courtesy Samsung Corp.)

- *Synchronous DRAM (SDRAM)*: SDRAM technology uses DRAM and adds a special interface for synchronization. It can run at much higher clock speeds than DRAM. SDRAM uses two independent memory banks. While one bank is recharging, the CPU can read and write to the other bank. Figure 7.4 shows a block diagram of SDRAM.
- *Extended Data Out RAM (EDORAM)* transfers blocks of data to or from memory.
- *Double Data Rate SDRAM (DDR SDRAM)* is a type of SDRAM that transfers data at both the rising edge and the falling edge of the clock. It can move data twice faster than SDRM; therefore, memory can run at the ½ clock rate. DDR2 and DDR3 are an advancement on the DDR technology and further increase the number of data transfers per clock cycle. DDR2 RAM provides 4 data transfers per cycle, and DDR3 transfers 8 data per clock cycle. For 100 MHz clock rate and 64 bits data bus, the transfer rates for DDR are

DDR = 100*2*8 = 1600 MB/s (MB/s).
DDR2 = 100*4*8 = 3200 MB/s.
DDR3 = 100*8*8 = 6400 MB/s.

- *Rambus DRAM (RDRAM)* was developed by Rambus Corporation. It uses multiple DRAM banks with a new interface that enables DRAM banks to transfer multiple words and also transfers data at the rising edge and the falling edge of clock. The RDRAM refreshing is done by the interface. The second generation of RDAM is called DRDRAM (Direct RDRAM), and it can transfer data at a rate of 1.6 Gbps. Figure 7.5 shows a RDRAM module.

## 7.2.2 DRAM Packaging

DRAM comes in different types of packaging such as SIMMs (single in-line memory module) and DIMM (dual in-line memory module).

Figure 7.6 shows SIMM, which is a small circuit board that one side of the board holds several chips. It has a 32 bit data bus.

DIMM is a circuit board that both sides of the board hold several memory chips but has a 64 bit data bus.

- *Static RAM (SRAM)* is used in cache memory. SRAM is almost 20 times faster than DRAM and is also much more expensive. Figure 7.7 shows diagram of 1 bit SRAM which is used for 6 MOSFET transistors.

## 7.2.3 ROM (Read-Only Memory)

Like its name suggests, information can be read only from *read-only memory (ROM)*. ROM holds information permanently, even while there is no power to the ROM; this type of memory is called *non-volatile memory*. Two types of ROM are listed below:

**Fig. 7.6** DRAM SIMM

**Fig. 7.7** 1 bit SRAM

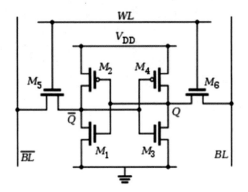

| Memory technology | Access time |
|---|---|
| SRAM | 0.5–2.5 ns |
| DRAM | 50–70 ns |
| Flash | $5 * 10^3 - 5 * 10^5$ |

**Table 7.2** Memory access time

- *Erasable Programmable Read-Only Memory (EPROM)*: EPROM can be erased with ultraviolet light and reprogrammed with a device called an EPROM programmer. Flash ROM is a type of EEPROM.
- *Electrically Erasable PROM (EEPROM)*: EEPROM can be erased by applying specific voltage to one of its pin and can be reprogrammed with an EPROM programmer.
- *Flash Memory*: flash memory is a non-volatile memory that has wide range of applications such as flash drive, solid-state drive, memory card, and embedded system. Flash memory is a type of EEPROM that allows multiple memory location to be written or erased on one operation. There are two types of technology use for flash memory, and they are NAND and NOR flash memories; NAND flash memory has smaller access time than NOR flash memory; most flash memory uses NAND technology.

### 7.2.4   Memory Access Time

The time the CPU places address on address bus and data appears on data bus or write the data into memory. Table 7.2 shows access time for different types of memory.

## 7.3   Hard Disk

Figure 7.8 shows the internal architecture of hard disk made of several platters, and platters hold the information; the functions of the heads are to read or write information from disk surface. The surface of platter is made of several tracks, and each track is divided into sectors as shown in Fig. 7.9.

### 7.3.1   Disk Characteristics

*Access Time*: the time that takes to start transfer data, and it is the sum of seek time and rotational delay.

*Seek Time*: the time that takes the head move to the proper track.

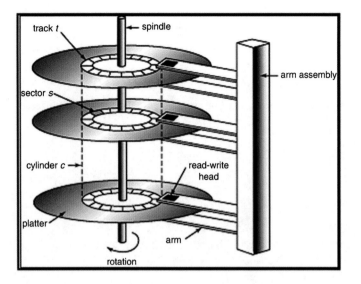

**Fig. 7.8**   Internal architecture of hard disk

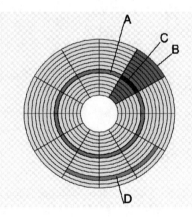

A : Track

B : Geometrical Sector

C: Track Sector

D: Cluster

**Fig. 7.9**   Surface of a platter

*Rotational Delay*: the time that it takes a sector to be positioned under read/write head and depend on rotation speed. The rotation speed represented by revolutions per minute (RPM) assumes the sector is away from head half of the track; therefore the rotation delay is calculated by

$$\text{Rotational delay} = \text{Time for half revolution} = 60 \ \text{s/RPM} * 2$$

*Disk Capacity*: capacity of a disk calculated by

| Table 7.3 Default cluster size | Disk size | NTFS cluster size |
|---|---|---|
| | 512–1024 MB | 1 kB |
| | 1024 MB–2 GB | 2 kB |
| | 2 GB–2 TB | 4 kB |

$$\text{Disk capacity} = \text{Number of surfaces}^*\text{Number of track/Surface}^*$$
$$\text{Number of sectors/Track}^*\text{Number of bytes/Sector}$$

### 7.3.2 Cluster

Each sector of a disk is 512 bytes (B), and cluster is made of one or more sectors, if a cluster is 1 kB, then it is made of two sectors. The 2 kB cluster is made of 4 sectors. Table 7.3 shows default values of cluster size.

**Example 7.1** A disk drive has 8 surfaces, each surface has 1024 tracks, each track has 64 sectors, and each sector can hold 512 B and rotation speed of 6000RPM.

(a) What is the capacity of disk?
(b) What is the rotational delay?

*Disk capacity* = 8 * 1024 * 64 * 512 = 268,435,456 B.
*Rotation delay* = 60/6000 * 2 = 0.005 s.

### 7.3.3 Disk File System

A file system defines organization of information stored in hard disk; the windows OS (operating system) offers FAT16 (file allocation table) and FAT32 which are used for early windows OS; currently most windows OS uses NTFS (New Technology File System). NTFS offers better security such as permission to restrict access and encryption.

## 7.4 Solid-State Drive (SSD)

Hard drive is a slow device, and it can be replaced by SSD. SSD is made of non-volatile NAND flash memory. Figure 7.10 shows architecture of SSD, and Table 7.4 compares SSD with HDD.

NAND Flash
Memory

Host Interface

Controller

**Fig. 7.10** Architecture of SSD

**Table 7.4** Comparing the SSD with HDD

| Characteristics | SSD | HDD |
| --- | --- | --- |
| Access time | 100 times faster than HDD | 5000–10,000 µs |
| Price | Expensive | Less expensive |
| Reliability | More reliable because it does not have any mechanical part | Less reliable |
| Capacity | Gigabytes | Terabytes |
| Power | Less power than HDD | More power than SSD |

**Fig. 7.11** Memory
hierarchy of a
microcomputer

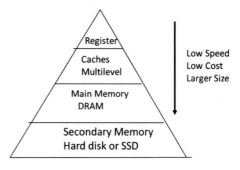

## 7.5 Memory Hierarchy

Computers come with four types of memory, which are arranged in a hierarchical fashion, as shown in Fig. 7.11:

**Table 7.5** Show price of different type of memory

| Memory type | SRAM | DRAM | SSD | HDD |
| --- | --- | --- | --- | --- |
| Cost | $8.00/MB | $0.16/MB | $0.20/GB | 0.05/GB |
| Access time | 0.5–2.5 ns | 50–70 ns | 70–150 ns | 5–20 ms |

1. *Register*: Register is fastest memory and it can read and write to it by single clock cycle.
2. *Cache memory*: Cache memory is the fastest type of memory and is most often used inside CPU called L1 cache, and it is faster than main memory and, therefore, more expensive than main memory.
3. *Main memory*: Main memory uses DRAM and SDRAM. The program to be executed moves from secondary memory (disk or tape) into main memory.
4. *Secondary memory*: Second memory refers to memory such as hard disk, SSD, and CD-ROM (Table 7.5).

## 7.5.1  Cache Memory

Each cache memory location is called cache line which can hold a block of data from main memory. For the most processors, the cache memory is located inside CPU and called L1 cache; there are two types of cache in a CPU:

1. Data cache (D-cache): Data cache holds the data and it can be read or write by CPU.
2. Instruction cache (I-Cache): Instruction cache holds instruction and CPU only read from I-cache.

## 7.5.2  Cache Terminology

*Miss*: when CPU accesses the cache and data is not in cache, it is called cache miss
*Hit*: when CPU accesses the cache and data is in the cache, then it is called cache hit
*Hit ratio*: number of hits/number of miss + number of hits (total number of reads)
*Block*: multiple of main memory locations is called block
*Physical address*: address generated by CPU to access main memory
*Virtual address*: address generated by CPU to access virtual memory or secondary memory
*Cache line or cache block*: each can line or block holds multiple byes or words; the size of cache line is the same block in main memory
*Temporal locality*: once a memory location referenced, then there is a high probability to be referenced again in near future

*Spatial locality*: when a memory location is accessed, then it is very likely the nearby
locations will be accessed soon.

### 7.5.3 Cache Memory Mapping Methods

Figure 7.12 shows a cache with 4 locations and main memory with 8 memory
locations, and the cache can hold only 4 memory locations of main memory; CPU
first accesses the cache if data is not in cache and then accesses main memory and
moves a block data into the cache; the question is where the data will store in the
cache; this bring subject of mapping methods.

The mapping methods are used to map a block of main memory into the cache
line (cache block), and the following are types of methods used for mapping:

1. Direct mapping
2. Associative mapping
3. Set associative

### 7.5.4 Direct Mapping

Figure 7.13 shows a cache memory with four cache lines, and each line holds 4 B;
the physical address seen by cache is divided into three fields and they are:

*Offset*: determines the size of cache line (number of bytes or words) since each cache
line can hold 4 memory locations and then the offset is 2 bits, and each block is
4 memory locations; also offset determines which of the four data to be transferred to CPU.
*Index*: index is the address to the cache; in this figure there are 4 cache lines and then
index is 2 bits.
*V-bit (valid bit)*: V-bit set to one to represent that data in cache line is valid.
*Tag*: Size of Tag = Size of Physical Address − (Size of Index + Size of Offset).

If physical address is 7 bits, then tag = 7 − (2 + 2) = 3 bits.

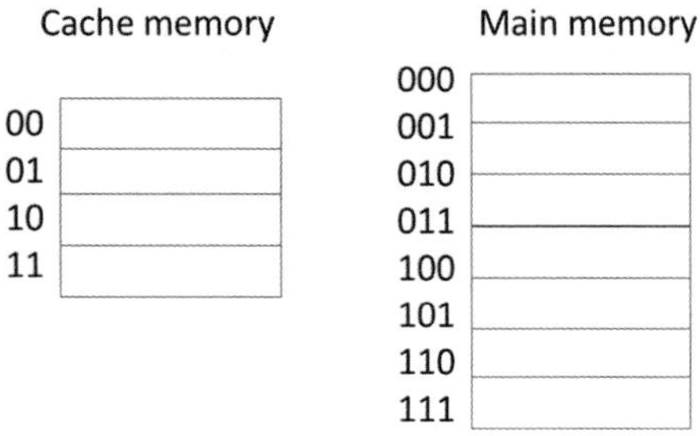

**Fig. 7.12** Cache and main
memory

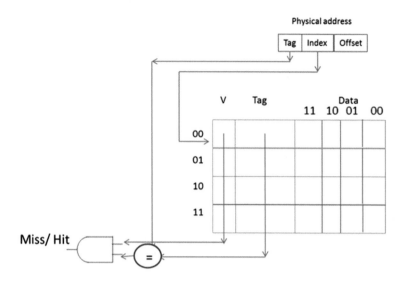

**Fig. 7.13** Cache memory with 4 cache lines

| Tag | Index | Offset |
|-----|-------|--------|
| 010 | 10 | 11 |

**Fig. 7.14** Format of physical address for 0101011

If valid bit is one and tag in address field matches with the tag stored in cache, then results are hit; otherwise is miss.

When CPU receives miss from cache, then access the main memory and transfer a block of data from main memory to cache line using the following equation for direct mapping:

$$\text{Cache line address} = (\text{Main Memory Block Number}) \text{Modulo N}$$

where N is the number of cache lines.

Consider Fig. 7.14 cache and assume CPU generates address 010 10 11, the Fig. 7.13 shows format of physical address seen by cache.

The CPU uses the index value to access cache line *10* and if the valid bit is zero results in a *miss*, the CPU accesses the main memory and transfers block *01010* to cache line *10* according to the following equation:

$$(01010)_2 = (10)_{10}$$

$$\text{Cache line number} = 10 \ \text{modulo} \ 4 = 2 \ \ or \ \ (10)_2$$

In this case each block is made up of 4 memory locations. Since the offset is two bits, it can refer any of 4 ($2^2$) chunks of data within a block of memory. The block at

**Fig. 7.15** Cache memory with four lines (four blocks)

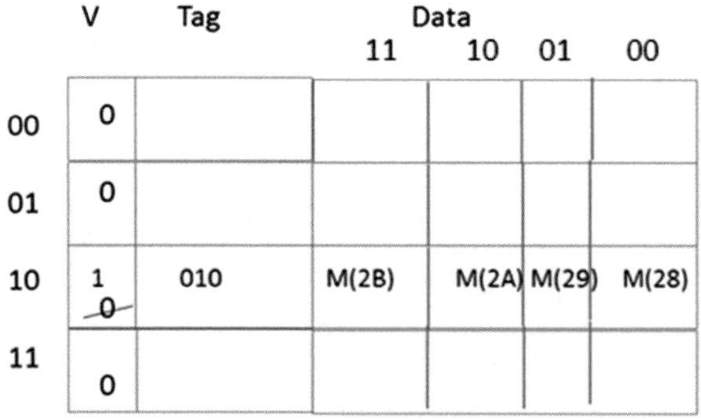

address *01010* in main memory with a 2 bit offset consists of memory locations 0101000 (M28), 0101001 (M29), 0101010 (M2A), and 0101011 (M2B).

In this case, M28, M29, M2A, and M2B are transferred to cache line *10*, and the V-bit is set to one. The tag is stored in the tag field of cache as shown in Fig. 7.15.

Now, if the CPU generates address *0101010*, where tag = *010*, index = *10*, and offset = *10*, the CPU uses the index to access cache line *10*. If on cache line *10* the V-bit equals *1* and the tag of the address matches with the tag in the cache line, then the result is a *hit*, and the CPU uses offset *10* to move data (M2A) in to the CPU. (Where offset *00* = M28, *01* = M29, *10* = M2A, *11* = M2B.)

**Example 7.2** Figure 7.16 shows the main memory and cache memory of a computer. CPU generates (in *hex*) addresses $0 \times 0$, $0 \times 2$, $0 \times 5$, and $0 \times 2$. Assuming the cache is empty at the beginning, show the contents of the cache.

In this example each block and cache line is 2 B and main memory consists of 8 blocks. Figure 7.17 shows the physical address as seen by the cache.

- The byte offset is one bit. (2 B per block).
- The index is two bits. (cache consist of 4 lines = $2^2$).
- The tag is one bit. (Bits in block − bits in index = 3−2 = 1).
- The CPU generates address $0 \times 0$ or *0000*. Therefore, the tag is *0*, the index is *00*, and the offset is *0*. The CPU accesses cache line *00* and the V-bit is zero which results in a *miss*. The CPU accesses main memory address *0000*, transfers block *000* to the cache line 00, and sets the tag bit to zero and the V-bit to one. Therefore, cache line *00* contains: V = *1*, Tag = *0*, Byte1 = $0 \times 5$, Byte0 = $0 \times 1$.
- Next the CPU generates address $0 \times 2$ or *0010*. The index is *01* so the CPU accesses cache line *01*, where the valid bit is zero which results in a *miss*. Then the CPU accesses main memory location *0010* and transfers block *001* into cache line *01*, changes the valid bit to one, and stores the tag part of the address into the cache line's tag. Therefore, cache line *01* is now: V = 1, Tag = 0, Byte1 = $0 \times 7$, Byte0 = $0 \times 6$.
- The CPU generates address $0 \times 5$ or *0101*. The index is *10*, so the CPU accesses cache line 10 and the result is a *miss*. The CPU accesses main memory location

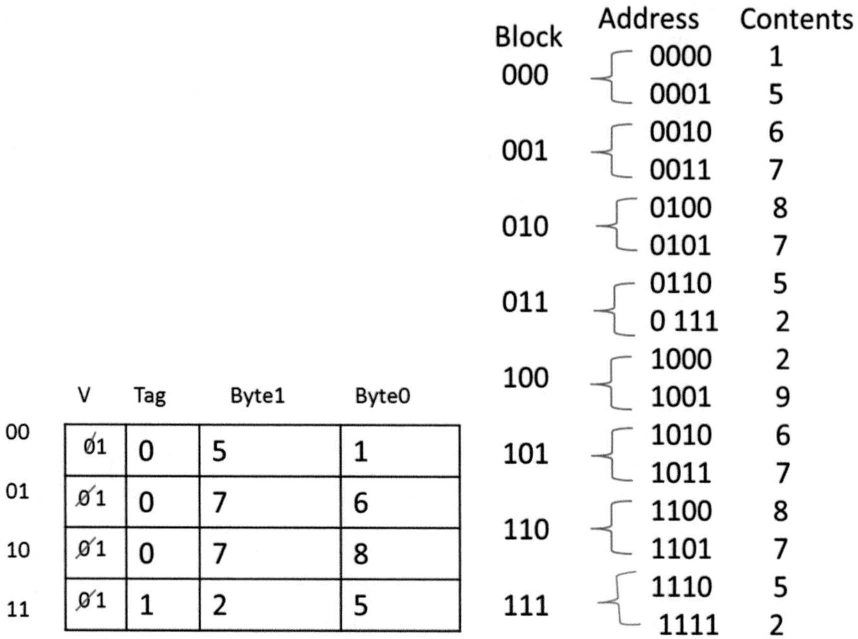

| Block | | Address | Contents |
|---|---|---|---|
| 000 | { | 0000 | 1 |
| | | 0001 | 5 |
| 001 | { | 0010 | 6 |
| | | 0011 | 7 |
| 010 | { | 0100 | 8 |
| | | 0101 | 7 |
| 011 | { | 0110 | 5 |
| | | 0 111 | 2 |
| 100 | { | 1000 | 2 |
| | | 1001 | 9 |
| 101 | { | 1010 | 6 |
| | | 1011 | 7 |
| 110 | { | 1100 | 8 |
| | | 1101 | 7 |
| 111 | { | 1110 | 5 |
| | | 1111 | 2 |

|    | V | Tag | Byte1 | Byte0 |
|----|---|-----|-------|-------|
| 00 | 0̷1 | 0 | 5 | 1 |
| 01 | 0̷1 | 0 | 7 | 6 |
| 10 | 0̷1 | 0 | 7 | 8 |
| 11 | 0̷1 | 1 | 2 | 5 |

**Fig. 7.16** Cache memory and main memory of Example 7.2

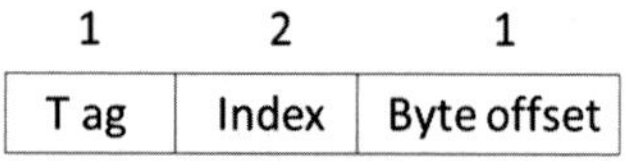

**Fig. 7.17** Physical address
format for Example 7.1

| 1 | 2 | 1 |
|---|---|---|
| Tag | Index | Byte offset |

*0101* and transfers block *010* to the cache, sets V to one, and stores the tag of the address into the cache's tag. Cache line *10* is now: $V = 1$, Tag $= 0$, Byte1 $= 0 \times 7$, Byte0 $= 0 \times 8$.

- The CPU generates address $0 \times 2$ or *0010*, so it accesses cache line *01*. The valid bit is 1 and the cache line tag matches the address tag, which results in a *hit*. The offset is *0*, so the CPU reads byte 0 from cache line *01*. This process will continue for other addresses.

Using the same size cache and main memory as above, blocks *000* and *100* will both be mapped onto cache line *00*. If the CPU generates addresses 0000, 1000, 0001, and 1001 consecutively, then results will be misses for all four addresses. In order to reduce misses, then the cache can be divided into sets, and mapping method is called set associative mapping.

### 7.5.5 Set Associative Mapping

In set associative mapping, the cache memory is divided into sets. The size of the set can vary: examples include two-way set associative, four-way set associative, and so on. Figure 7.18 shows a two-way set associative cache, and Fig. 7.19 shows the physical address format seen by the cache. Using the previous examples as a base, the byte offset will remain one bit since it refers to one of two bytes stored on the cache line. The *set* identifier will be one bit that defines the set address; the tag size is calculated as follows:

$$\text{Tag size} = \text{Physical address size} - \text{Set address} - \text{Offset}$$

Assume the physical address is 4 bits, then Fig. 7.19 shows the physical address seen by cache.

**Example 7.3** Consider the main memory from Fig. 7.16, where the cache is empty and it is divided into two sets as shown in Fig. 7.20. Show the contents of the cache if the CPU generates addresses $0 \times 0$, $0 \times 8$, $0 \times 0$, and $0 \times 8$.

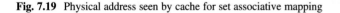

Tag Size = Physical address size – Set address – Offset

| V | Tag | Byte1 | Byte0 |   | V | Tag | Byte1 | Byte0 |
|---|-----|-------|-------|---|---|-----|-------|-------|
| 0 | 0 |  |  |  | 0 | 0 |  |  |  |
| 1 | 0 |  |  |  |   | 0 |  |  |  |

**Fig. 7.18** Two-way set associate cache

| 2 | 1 | 1 |
|---|---|---|
| Tag | Set | Byte Offset |

**Fig. 7.19** Physical address seen by cache for set associative mapping

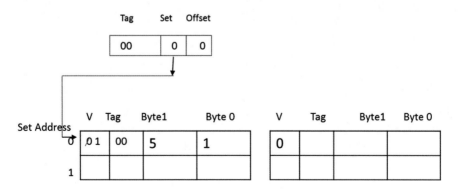

**Fig. 7.20** Content of cache for address 0000

- The CPU generates address $0 \times 0$ or *0000* and accesses cache set *0*. Both cache lines in the set 0 have valid bits of *0*, so the result is a *miss*. The CPU then accesses main memory, transfers contents of memory locations *0000* and *0001* into the cache, then changes the valid bit to one, and stores the tag part of the address into the cache's tag.

Next, the CPU generates address $0 \times 8$ or *1000* and accesses cache set *0*. First line of the cache has a valid bit of 1 but the tag does not match (*10* vs. *00*). The CPU then accesses main memory, transfers contents of memory locations *1000* and *1001* into second line of the cache in the set 0, and changes the valid bit to one and stores the tag part of the address into the cache's tag. Now, when the CPU generates the addresses $0 \times 0$ and $0 \times 8$ again result *hits*.

## 7.5.6  Replacement Method

In set associative, when CPU brings a new block into cache, then one of the cache lines must be replaced with new block; consider Fig. 7.21. If CPU generates address 0100 (4) (tag = 01, set address = 0, and offset is 0), it accesses set 0, and both cache lines have valid bit one, but tags in cache lines do not match with address tag results miss, and then CPU accesses main memory and must move the contents of memory locations 0100 and 0101 into cache. CPU uses *least recently used (LRU)* method which moves new block from main memory and replaces it with block that has been longer in the cache. This method accomplishes by adding LRU bit to each cache line of the cache as shown in Fig. 7.22.

In Fig. 7.22, assume both cache lines in set 0 are empty, a new block moves into the first cache line in set 0, and LRU changed from 0 to 1; the second block moves to second cache line of set 0, and LRU changed from 0 to 1, but at the same time, LRU of first cache line will change from 1 to 0; therefore, the cache line with LRU = 0 contains block that has been longer in cache line.

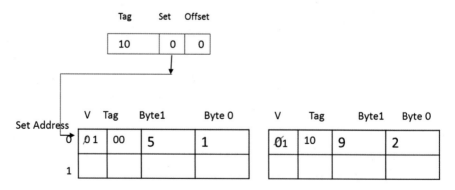

**Fig. 7.21** Contents of cache for addresses 0000 and 1000

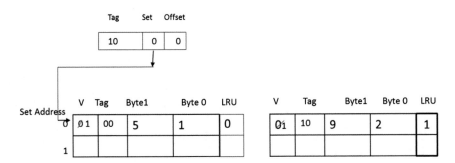

**Fig. 7.22** Two set associative with LUR

**Fig. 7.23** Associative mapping

| Valid | Address | Byte1 | Byte0 |
|-------|---------|-------|-------|
| 1     | 0 00    | 5     | 1     |
| 1     | 1 00    | 9     | 2     |
| 0     |         |       |       |
| 0     |         |       |       |

## 7.5.7 Fully Associative Mapping

In fully associative mapping, the entire address is stored in the cache with its data. Figure 7.23 shows a fully associative cache with four lines after the CPU has accessed main memory locations $0 \times 0$ and $0 \times 8$. If the CPU next generates address $0 \times 0$ or *0000*, then it will compare the address with each address in the cache, and if it matches, then it will read the data from the cache.

## 7.5.8 Cache Update Methods

1. *Write Through*: When new information is written to the cache, main memory is also updated.
2. *Buffered Write Through*: There is a buffer between cache and main memory, when new information is written to the cache, this information is written into buffer, and the CPU can access this memory before the new information can be written into main memory.
3. *Write Back (Copy Back)*: In this method, only the cache is updated, and main memory will be updated when the corresponding cache line is overwritten. In this method each cache line has a *dirty bit* to indicate if cache line has been modified or not.

### 7.5.9  Effective Access Time (EAT) of Memory

The performance of memory depends on hit ratio of cache, the EAT of memory is calculated by

$$EAT = H * Tc + (1 - H) * Tm$$

The H is hit ratio of the cache, Tc is the cache access time, (1-H) is miss rate, and Tm is memory access time.

**Example 7.4**  A computer has a cache memory with access 10 ns and main memory access time 100n, a program was executed and result 90% hit ration. What is EAT of the memory

$$EAT = 0.9 * 10 + (1 - 0.9)200 = 29 \text{ ns}$$

### 7.5.10  Virtual Memory

Virtual memory is a HDD or SSD; it is used to store application data and instruction that is currently not needed to be process by the CPU. Virtual memory enables a system to run application larger than main memory. Disk is seen by CPU as virtual memory, if CPU has 16 address lines, then the size of virtual memory will be $2^{16}$ B. Application resides in disk and it is called process. When user runs a program, the operating system moves the pages of process into main memory. Virtual memory is divided into the pages as shown in Fig. 7.24; the process A occupied pages P0 through P4, and process B occupied pages P5–P9.

The CPU generates a virtual address (to access an address in the disk) of $V$-bits. These bits are divided into two identifiers: a virtual page number of $M$ bits and a page offset of $N$ bits as shown in Fig. 7.25. The total number of pages in the system is equal to $2^M$, and the number of bytes (in a byte-addressable system) in a page is equal to $2^N$. The total number of addresses in a system is equal to $2^{(N + M)}$ or $2^V$.

**Example 7.4**  The capacity of a virtual disk is 2 MB (megabytes), and each page is 2 kB (kilobytes) in a byte-addressable system.

(a) What are values of N and M?
(b) How many pages are in the disk?

Since each page is 2 kB, $2^N = 2048$ B, meaning N equals 11.
The capacity of the disk is 2 MB. $2^V = 2$ M equals $2^{21}$ so V = 21 bits.
The number of pages, then, is equal to $2^{(21-11)}$. The disk contains $2^{10}$ or 1024 pages, and the size of each page is 2 kB.

**Fig. 7.24** Virtual memory

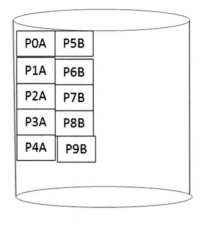

**Fig. 7.25** Virtual address
format

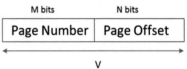

### 7.5.10.1   Page Table

With main memory divided into blocks, the size of each block (or frame) is equal to
the page size. When the CPU transfers a page into main memory, it records the page
number and corresponding block in the *page table*. The address line of the page table
is the page number. Each line contains the frame or block number of the matching
location in main memory and a valid bit that indicates whether the line is valid or not.
Figure 7.26 shows a page table wherein pages P0, P1, P3, and P4 are transferred to
the blocks 2, 3, 0, and 1, respectively.

   Each process has its own page table stored in main memory. Since cache memory
is faster, part of the page table referred to as the *translation lookaside buffer* (TLB) is
stored in the cache. The TLB uses associative mapping.

## 7.5.11   Memory Organization of a Computer

Figure 7.27 shows memory organization of a computer which in this example
consists of:

(a) Virtual memory (hard disk or solid-state drive).
(b) Main memory (A type of DRAM).
(c) Cache memory (SRAM).
(d) A page table which keeps track of pages in main memory.
(e) The TLB which holds a part of the page table.

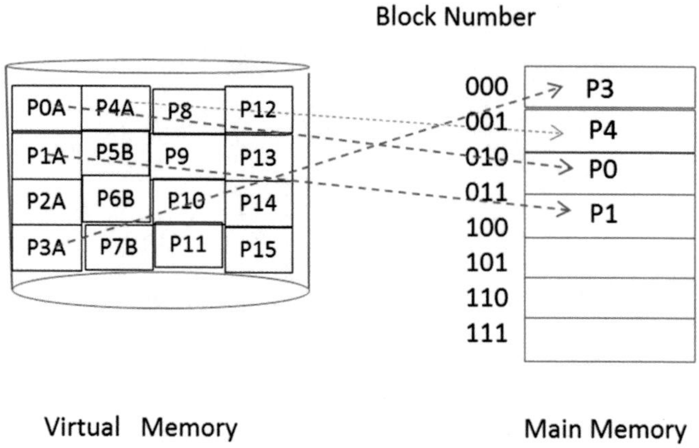

Block Number

Virtual  Memory                                    Main Memory

| Page Number | V | Block Number |
|---|---|---|
| 0000 | 1 | 010 |
| 0001 | 1 | 011 |
| 0010 | 0 | |
| 0011 | 1 | 000 |
| 0100 | 1 | 001 |
| 0101 | 0 | |
| 0110 | 0 | |
| 0111 | 0 | |
| 1000 | 0 | |
| 1001 | 0 | |
| 1010 | 0 | |
| 1011 | 0 | |
| 1100 | 0 | |
| 1101 | 0 | |
| 1110 | 0 | |
| 1111 | 0 | |

**Fig. 7.26** Virtual memory, memory, and page table

### 7.5.11.1  Memory Operation

The following steps describe the operation of memory of a computer. First, the CPU generates a virtual address and checks the TLB to see if the corresponding page is in main memory already or not.

(a) If the TLB indicates that the corresponding page is in memory, then generate a physical address and check if the data is in the cache.

   1. If it is in the cache, then this called a *hit* and it reads the data from the cache.
   2. If it is not in the cache, then it is called a *miss*, and the CPU accesses memory and moves a block of data into the cache then reads it.

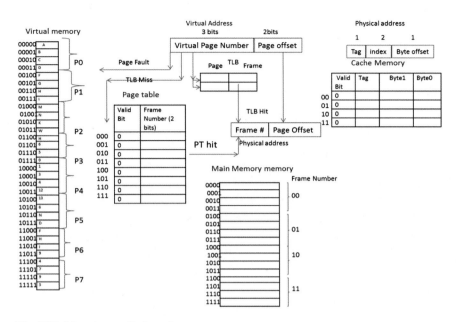

**Fig. 7.27** Memory organization of a computer

(b) If the corresponding page is not in the TLB, then the CPU checks the page table.

1. If the corresponding page is in main memory, then update the TLB and repeat from step 1.
2. If corresponding page is *not* in main memory, then move the page from virtual memory into main memory, update the page table, update the TLB, and repeat from step 1.

(c) If the corresponding page is not in the page table, then the CPU moves the page from virtual memory into main memory, updates the page table, updates the TLB, and repeats from step 1.

## Questions and Problems

1. Distinguish between volatile and non-volatile memory.
2. What does RAM stand for?
3. List three different types of RAM.
4. Which of the following memory types are used for main memory?

(a) ROM and SDRAM.
(b) SRAM and DRAM.
(c) SDRAM and DRAM.
(d) DRAM and EPROM.

5. _____ holds information permanently, even when there is no power.

   (a) ROM.
   (b) DRAM.
   (c) RAM.
   (d) SRAM.

6. What does ROM stand for?
7. Is flash memory a type of RAM or ROM?
8. What is the difference between EEPROM and EPROM?
9. What is capacity of a memory with 10 address lines and holds one byte per memory location?
10. What is the primary application of SRAM?
11. What is the primary application of DRAM?
12. Define the following terms:

    (a) Track.
    (b) Sector.
    (c) Cluster.

13. A hard disk consists of 4 surfaces, each surface consists of 80 tracks, and each track consists 32 sectors. Each sector holds 512 B. What is the capacity of this disk?
14. What is the function of file allocation table (FAT)?
15. List types of memory in a computer from fastest to slowest.
16. What are the types of cache?
17. What type of memory is used for cache memory?
18. What is virtual memory?
19. Distinguish between a virtual address and physical address.
20. Physical address determines size of

    (a) Virtual memory
    (b) Physical memory
    (c) Cache memory

21. Show the format of a virtual address.
22. What is hit ratio?
23. Explain temporal locality.
24. Explain spatial locality.
25. List cache mapping methods.
26. Show format of address seen by the cache for direct mapping.
27. List cache mapping methods.
28. Show format of address seen by the cache for set associative mapping.
29. What is the function of a page number in a virtual address?
30. How many bits is the page offset if each page holds 8 kB?
31. What is the function of the page table?
32. What information is stored in TLB? Where is the TLB stored?
33. List cache mapping methods.

34. What is the advantage of set associative versus direct mapping of caches?
35. What are the three write policies used for memory?
36. _____ is the fastest type of memory.

    (a) Cache memory
    (b) Main memory
    (c) Secondary memory
    (d) Hard disk

## Problems

1. The following main and cache memory are given. The CPU generates addresses $0 \times 0, 0 \times 2, 0 \times 3, 0 \times 4, 0 \times 5, 0 \times 3, 0 \times 6, 0 \times 6, 0 \times 7, 0 \times B, 0 \times D$, and $0 \times F$. Show the contents of the cache and find the hit ratio.

| ADDRESS | Contents |
|---------|----------|
| 0000 | 5 |
| 0001 | 0 |
| 0010 | 1 |
| 0011 | 11 |
| 0100 | 15 |
| 0101 | 09 |
| 0110 | 16 |
| 0111 | 23 |
| 1000 | 65 |
| 1001 | 01 |
| 1010 | 8 |
| 1011 | 9 |
| 1100 | 15 |
| 1101 | 0 |
| 1110 | 2 |
| 1111 | 5 |

| V | Tag | Data |
|---|-----|------|
|   |     |      |
|   |     |      |
|   |     |      |
|   |     |      |
|   |     |      |
|   |     |      |
|   |     |      |

2. The Ffollowing Mmemory and cache are given, if CPU generates address 0, 1, 2, 3, 8,9, 11, 13, 1F and 1E in hex

    (a) Show the contents of the cache using direct mapping
    (b) What is the hit ratio?

## Memory

| Address | Content | Address | Content |
|---|---|---|---|
| 00000 | 5 | 10000 | 5 |
| 00001 | 3 | 10001 | 0 |
| 00010 | 11 | 10010 | 1 |
| 00011 | 6 | 10011 | 11 |
| 00100 | 7 | 10100 | 15 |
| 00101 | 8 | 10101 | 09 |
| 00110 | 9 | 10110 | 12 |
| 00111 | 12 | 10111 | 23 |
| 01000 | 0 | 11000 | 65 |
| 01001 | 0 | 11001 | 21 |
| 01010 | 8 | 11010 | 8 |
| 01011 | 7 | 11011 | 7 |
| 01100 | 9 | 11100 | 9 |
| 01101 | 0 | 11101 | 0 |
| 01110 | 2 | 11110 | 2 |
| 01111 | 5 | 11111 | 5 |

## Cache Memory

|    | V | Tag | Data | Data |
|----|---|-----|------|------|
| 00 |   |     |      |      |
| 01 |   |     |      |      |
| 10 |   |     |      |      |
| 11 |   |     |      |      |

3. Using information of problem#2 with the following cache

   (a) Show the contents of the cache
   (b) What is the hit Rate?

|   | V | Tag | Data | Data | Data | Data |
|---|---|-----|------|------|------|------|
| 0 |   |     |      |      |      |      |
| 1 |   |     |      |      |      |      |

4. The following memory and cache memory is given. CPU generates addresses $0 \times 1, 0 \times 2, 0 \times 1, 0 \times 8, 0 \times 9, 0 \times 1C, 0 \times 1D, 0 \times 3$, and $0 \times 4$.

   (a) Show the contents of the cache using two-way set associative mapping; assume a LRU replacement policy.
   (b) What is the hit rate?

| Set Address | V | Tag | B1 | B0 | LRU | V | Tag | B1 | B0 | LRU |
|---|---|---|---|---|---|---|---|---|---|---|
| 00 | 0 |  |  |  | 0 | 0 |  |  |  | 0 |
| 01 | 0 |  |  |  | 0 | 0 |  |  |  | 0 |
| 10 | 0 |  |  |  | 0 | 0 |  |  |  | 0 |
| 11 | 0 |  |  |  | 0 | 0 |  |  |  | 0 |

| Address | Content |
|---------|---------|
| 00000 | 5 |
| 00001 | 3 |
| 00010 | 11 |
| 00011 | 6 |
| 00100 | 7 |
| 00101 | 8 |
| 00110 | 9 |
| 00111 | 12 |
| 01000 | 0 |
| 01001 | 0 |
| 01010 | 8 |
| 01011 | 7 |
| 01100 | 9 |
| 01101 | 0 |
| 01110 | 2 |
| 01111 | 5 |

| Address | Content |
|---------|---------|
| 10000 | 5 |
| 10001 | 0 |
| 10010 | 1 |
| 10011 | 11 |
| 10100 | 15 |
| 10101 | 09 |
| 10110 | 12 |
| 10111 | 23 |
| 11000 | 65 |
| 11001 | 21 |
| 11010 | 8 |
| 11011 | 7 |
| 11100 | 9 |
| 11101 | 0 |
| 11110 | 2 |
| 11111 | 5 |

5. Assume a computer has 1 M bytes main memory and a cache of 64 blocks, where each cache block contains 8 bytes

   (a) What is the format of address seen by cache?
   (b) How many blocks are in main memory?
   (c) Which block does CPU access with address 0AF56

6. Assume a computer has 512 M byes of memory with 1024 blocks of cache and each cache block holds 16 bytes

   (a) How many blocks are in main memory?
   (b) What is the format of address seen by cache using direct mapping?
   (c) What is the format of address seen by cache using 2-ways set associative?

7. A computer has 24 bit physical addresses and each memory location holds one byte. This computer has 64 cache lines and each line holds 16 Bytes. Show the format of the address (tag, index, and byte offset) using

   (a) Direct mapping
   (b) 4-way set associative
   (c) 8-way set associative

8. A computer has 32 kB of virtual memory and 8 kB of main memory with a page size of 512 B.

   (a) How many bits are in the virtual address?
   (b) How many pages are in virtual memory?

     (c) How many bits are required for the physical address?

     (d) How many frames or blocks are in main memory?

9. A computer with 256 MB of virtual memory, 4 MB of main memory, and 8 kB of cache memory. Assume a page size of 2 kB.

     (a) What is the size of a virtual address?

     (b) What is the size of a physical address?

     (c) How many pages are in virtual memory?

     (d) How many blocks are in main memory?

     (e) What is the size of the page table? (Include the number of locations and the total size of each location including all information.)

10. A Computer Has 20 Bits of Virtual memory and each Page Is 2 kB.

     (a) What is the size of virtual memory?

     (b) How many pages are in virtual memory?

11. A computer with 4 words per block has 4 K blocks of cache and 1 M blocks of main memory.

     (a) What is the size of a physical address?

     (b) Determine the size of the tag, index, and word offset of physical address using direct mapping.

     (c) Determine the size of the tag, set, and word offset of physical address using Two-way set associative mapping.

12. CPU of Fig. 7.26 generates addresses $0 \times 00$ and $0 \times 0b$; assume page 0 map into block 1 and page 2 map in block 0, show the contents of page table.

13. A computer has 64 M bytes of virtual memory and 16 M bytes of main memory, assume each page is 4 Kbytes.

     (a) What is size of virtual address?

     (b) How many bits is Physical address?

     (c) Show address field of Virtual address (Page Number and Page Offset)?

     (d) How many blocks are in main memory?

     (e) Calculate the physical address for Virtual address 005444, assume page 5 maps to block 1.

14. Use Fig. 7.27 and assume CPU generates addresses 02, 03, 1F, 1E, and 19 in Hex, assume page 0 maps into frame# 3, page 7 maps into frame #2, and page6 maps to frame# 0,

     (a) Show the content of page table, TLB, memory, and cache memory.

          Address in Hex TLB Hit/Miss Page Fault (yes/No) Physical address cache Hit/Miss.

          02

          03

          1F

1E
19

15. A computer has 1 MB bytes of virtual memory, 512 k bytes of main memory, and page size of 32 k bytes. The following figure shows TLB and page table of the computer.

(a) How many bits are in virtual address?
(b) How many pages are in virtual memory?
(c) How many bits require for physical address?
(d) How many frames are in main memory?
(e) The following virtual addresses are given in Hex, find physical addresses and identify if each address generates page fault or not

- 38AAA
- 47BBB
- 19EEE

**Page Table**

| | V | Frame# |
|---|---|---|
| 00000 | 1 | 1000 |
| 00001 | 0 | |
| 00010 | 0 | |
| 00011 | 1 | 1111 |
| 00100 | 0 | |
| 00101 | 0 | |
| 00110 | 0 | |
| 00111 | 1 | 0100 |
| 01000 | 1 | 0101 |

**TLB**

| Page # | Frame # |
|---|---|
| 00111 | 0100 |
| 01000 | 0101 |

# Chapter 8
# Assembly Language and ARM Instructions Part I

**Objectives: After Completing this Chapter, you Should be Able to:**
- Explain the function of compiler and the assembler.
- Convert HLL to the machine language.
- Show ARM processor architecture.
- Describe the function of processor state register (PSR).
- List instruction classification based on number of the operands.
- Learn different types of the ARM instructions.
- Describe the operation of conditional instructions.
- Convert HLL to the assembly language.
- Explain the shift and the rotate instructions.
- Explain the operation of stack instructions.
- Explain application of the Branch instructions.

## 8.1  Introduction

Programmers use high-level language to develop application program; in order for the program to become an executable form, it must be converted in machine code (binary).

Figure 8.1 shows a high-level language (HLL) converted to machine code, the compiler converts HLL into assembly language, and then assembler converts assembly language to machine language (bits) by assembler.

Each CPU has a set of instructions which represents the type of operations the CPU can perform, and these instructions are represented in *mnemonic* forms or abbreviation, for example, the addition instruction is represented by "ADD," and subtraction instruction is represented by "SUB."

ADD R1, R2, R3 means add contents of R2 with R3 register and store results in R1 register. R1, R2, and R3 are called operands.

© The Author(s), under exclusive license to Springer Nature Switzerland AG 2022
A. Elahi, *Computer Systems*, https://doi.org/10.1007/978-3-030-93449-1_8

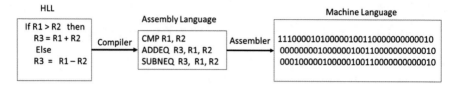

**Fig. 8.1**  Converting HLL to machine language

The following HLL are converted to assembly language:

**HLL Assembly Language**
R3=R1+R2 ADD R3, R1, R2
R3=R1-R2 SUB R3, R1, R2

The programmer uses instructions to write assembly language. The applications of assembly language are as follows:

• Assembly language is used for writing fastest code.
• It helps to better understand HLL.
• Writing compiler for HLL requires knowledge of assembly language.
• It is used in embedded system and driver.
• HLL may not provide access to hardware then assembly language can be used.

## 8.2  Instruction Set Architecture (ISA)

Manufacturers of CPUs publish a document that contains information about the processor such as list of registers, function of each register, size of data bus, size of address bus, and list of instructions that can be executed by the CPU. Each CPU has a known instruction set that a programmer can use to write assembly language program. Instruction sets are specific to each type of processor. Pentium processors use a different instruction set than ARM processor. The instructions classified are based on number of operands or type of operation.

### 8.2.1  Classification of Instruction Based on Number of Operands

#### 8.2.1.1  No Operand Instructions

The following are some of the instructions that do not require any operands:

HLT                Halt the CPU
NOP                No operation
PUSH *operand*:    Push operand into top of the stack
POP *operand*      Remove the operand from top of the stack

### 8.2.1.2   One-Operand Instructions

The following are some of the instructions that require one operand.

INC   *operand*   Example: INC R1—Increment register R1 by 1
DEC   *operand*   Example: DEC R1—Decrement register R1 by 1
J     *target*    Jump to memory location labeled by target
ADD   *operand*   Add operand to the accumulator (ACC) ACC → ACC + *operand*

### 8.2.1.3   Two-Operand Instructions

The following are some of the instructions that require two operands.

```
ADD Rd, Rn Example: ADD R1, R2 - R1←R1+R2
```

Intel Instruction Set Architecture uses two operands.

```
MOV EAX, EBX ; EAX ← EBX
```

### 8.2.1.4   Three-Operand Instructions

Most modern processors use instructions with three operands, such ARM, MIPS, and Itanium

```
ADD R1, R2, R3 ; R1← R2 +R3
```

## 8.3   ARM Processor Architecture

Advanced RISC Machine (ARM) was developed by the Acorn Company. ARM is a leader supplier of microprocessors in the world, ARM develops the core CPU, and thousands of suppliers add more functional units to the core. ARM processor is one of a family of CPUs based on the **RISC** (reduced instruction set computer), ARM offers 32 bit, 64 bit, 128 bits processor. All ARM processors use the same instruction set and they are targeted for different applications such as

Cortex-A: Use for performance and optimal power
Cortex-M: Use for most energy efficient embedded devices
Cortex-R: Use for real-time performance
Ethos – NPUs: Use for machine learning
SecurCore: Use for security application
Neoverse: Use for cloud computing

ARM uses two types of instruction called Thumb and Thumb-2. Thumb instructions are 16 bits and thumb-2 instructions are 32 bits; currently most ARM processors use 32-bit instructions.

ARM contains 15 registers called R0 through R15, R0 through R12 called general propose registers. ARM is able to execute Thumb instructions (16-bit instructions) and Thumb-2 32 bits instruction. Thumb instructions use registers R0 through R7.

ARM is intended for applications that require power efficient processors, such as telecommunications, data communication (protocol converter), portable instrument, portable computer, and smart card. ARM is basically a 32-bit RISC processor (32-bit data bus and address bus) with fast interrupt response for the use in real-time applications. A block diagram of ARM7 processor is shown in Fig. 8.2.

### 8.3.1  Instruction Decoder and Logic Control

The function of instruction decoder and logic control is to decode instructions and generate control signals to other parts of processor for execution of instructions.

### 8.3.2  Address Register

To hold a 32-bit address for address bus.

### 8.3.3  Address Increment

It is used to increment an address by four and place it in address register.

### 8.3.4  Register Bank

Register bank contains 31 32-bit registers and 6 status registers.

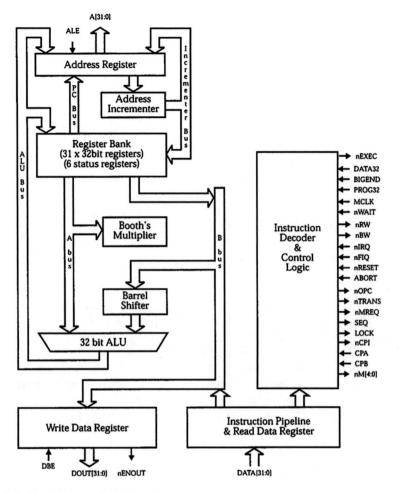

**Fig. 8.2** Block diagram of ARM7 architecture

## 8.3.5   *Barrel Shifter*

It is used for fast shift operation.

## 8.3.6   *ALU*

32-bit ALU is used for arithmetic and logic operation.

### 8.3.7   Write Data Register

The processor put the data in Write Data Register for write operation.

### 8.3.8   Read Data Register

When processor reads from memory it places the result in this register.

### 8.3.9   ARM Operation Mode

ARM can operate in one of the following modes:

1. *User mode*: Use for normal operation.
2. *IRQ mode*: This interrupt mode is designed for handling interrupt operations.
3. *Supervisory mode*: Used by operating system.
4. *FIQ mode*: Fast interrupt mode.
5. *Undefined mode*: When an undefined instruction executed.
6. *Abort mode*: This mode indicates that current memory access cannot be completed, such as when data is not in memory and the processor requires more time to access disk and transfer block of data to memory.

## 8.4   ARM Registers

ARM7 has 31 general registers and 6 status registers. At user mode, only 16 registers and 1 Program Status Register (PSR) are available to programmers. The registers are labeled R0 through R15. R15 is used for program counter (PC), R14 is used for link register, and R13 is used for stack pointer (SP). Figure 8.3 shows user mode registers.

### 8.4.1   Current Program Status Register (CPSR)

Figure 8.4 shows the format of PSR. This register is used to store control bits and flag bits. The flag bits are N, Z, C, and V, and the control bits are I, F, and M0 through M4. The flag bits may be changed during a logical, arithmetic, and compare operation.

**Fig. 8.3** User mode
registers

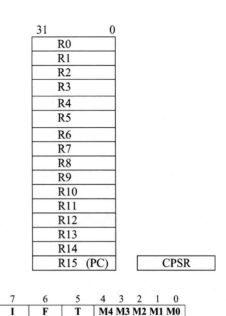

31                                           0

| R0 |
| R1 |
| R2 |
| R3 |
| R4 |
| R5 |
| R6 |
| R7 |
| R8 |
| R9 |
| R10 |
| R11 |
| R12 |
| R13 |
| R14 |
| R15 (PC) |

CPSR

| 31 | 30 | 29 | 28 | 27 | 7 | 6 | 5 | 4 3 2 1 0 |
|----|----|----|----|--------|---|---|---|-------------|
| N | Z | C | V | Unused | I | F | T | M4 M3 M2 M1 M0 |

**Fig. 8.4** Storage format for CPSR

## 8.4.2  *Flag Bits*

*N (negative)*: $N = 1$ means result of an operation is negative, and $N = 0$ means result
 of an operation is positive.
*Z (zero)*: $Z = 1$ means result of an operation is zero, and $Z = 0$ result of an operation
 is not zero.
*C (carry)*: $C = 1$ means result of an operation generated a carry, and $C = 0$ means
 result of an operation did not produce a carry.
*V (overflow)*: $V = 1$ means result of an operation generated an overflow, and $V = 0$
 means result of an operation did not generate an overflow.

## 8.4.3  *Control Bits*

*I (interrupt bit)*: When this bit sets to one, it will disable the interrupt, and this means
 that the processor does not accept any software interrupt.
*F*-bit is used to disable and enable *fast interrupt request mode* (FIQ) mode.
*M4, M3, M2, M1, and M0* are mode bits, and they are equal to 10,000 for user mode.
*T (State bit)*: $T = 1$ Processor executing Thumb instructions, $T = 0$ processor
 executing ARM instructions.

## 8.5   ARM Instructions

ARM architecture supports Thumb 16-bit and Thumb-2 32-bit instruction set. Most of the ARM instructions use three operands. These instructions are classified based on their instructions format and operations that are listed as follows:

(a) Data processing instructions
(b) Single data swap
(c) Shift and rotate instructions
(d) Unconditional instructions and conditional instructions
(e) Stack operations
(f) Branch
(g) Multiply instructions
(h) Data transfer

### *8.5.1   Data Processing Instructions*

The data processing instructions are as follows: AND, EOR, SUB, RSB, ADD, ADC, SBC, RSB TST, TEQ, CMP, CMN, ORR, MOV, BIC, and MNW. Data processing instructions use register operands and immediate operand. The general format of data processing instructions is as follows:

**Mnemonic {S}{Condition} Rd, Rn, operand2**
**Mnemonic**: Mnemonic is abbreviation of an operation such as ADD for addition
**{}**: Commands inside the { } is optional such as S and condition
**S:** When an instruction contains S mean update the Processor Status Register (PSR) flag bits
**Condition**: Condition define the instruction will execute if meet the condition
**Rd**: Rd is destination register
**Rn**: Rn is operand1
**Operand2**: Operand2 can be register or immediate value

#### 8.5.1.1   Registers Operands

The operands are in registers. First register is destination register, second register is operand1, and third register is operand2.

The following are arithmetic and logic operations instructions with register operands.

***ADD R0, R1, R2*** ;R0=R1+R2 Add contents of register R1 with register R2 and place the result in register R0.
***ADC R0, R1, R2;*** ;R0 = R1+R2 +C Add with carry C is carry bit.

*SUB R0, R2, R3* ;R0=R2-R3 where R2 is first operand and R3 is second operand
*SBC R0, R2, R3;* ;R0=R2-R3+C-1 SUB with carry.
*RSB R0, R2, R5* ;R0= R5-R2 Reverse SUB.
*RSC R0, R2, R5* ;R0=R5-R2+C-1 Reverse sub with carry.
*AND R0, R3, R5* ;R0= R3 AND R5.
*ORR R7, R3, R5;* ;R7=R3 OR R5.
*EOR R0, R1, R2* ;R0 = R1 Exclusive OR with R2.
*BIC R0, R1, R2* ;Bit clear. The one in second operand clears corresponding
bit in first operand and stores the results in destination register.

**Example 8.1** Assume contents of R1 is 1111111111011111, and R2 is 1000 0100 1110 0011 after execution of *BIC R0,R1, R2* the R0 contains 0111 101100011100.

### 8.5.1.2  Immediate Operand

In immediate operand, operand2 is an immediate value, and maximum can be 12 bits

*ADD R1, R2, #&25* ;R1=R2+&25, # means immediate and & means the immediate
value is in hexadecimal.
*AND R2, R3, #&45* ;R2 = R3 AND &45.
*EOR R2, R3, #&45* ;R2= R3 Exclusive OR &45.

**Example 8.2**  What are the contents of R1 after executing the following instruction? Assume R2 contains 0x12345678

```
ADD R1, R2, #0x345
```

The ADD instruction will add contains of R2 with 0x2345 and store the result in R1, then R1 = 0x123459BD.

#### Setting Flag Bits of PSR

The above instructions do not affect the flag bit of PSR because the instructions do not have option S. By adding suffix S to the instruction, the instruction would affect the flag bit.

*ADDS R1, R2, R3* ;The suffix S means set appropriate flag bit.
*SUBS R1, R2, R2;* ;This will set zero flag to 1.

## 8.5.2  Compare and Test Instructions

ARM processor uses the compare and test instructions to set flag bits of PSR and the following are compare and test instructions.

**CMP, CMN, TST, and TEQ,** these instruction use two operands for compare and test, the result of their operations do not write to any register.

### 8.5.2.1   CMP Instruction (Compare Instruction)

The CMP instruction has the following format:

```
CMP Operand1, Operand2
```

The CMP instruction compares Operand1 with Operand2; this instruction subtracts Operand2 from Operand1 and sets the appropriate flag. The flag bit set based on the result of the operation as follows:

**Z** flag set if Operand2 equal operand1
**N** flag is set if operand1 less than operand2
**C** flag is set if result of operation generate carry

**Example 8.4**  Assume R1 contains 0x00000024, and R2 contains 0x00000078; the operation CMP R1, R2 will set N flag to 1.
    **CMP Rd, immediate value**, the immediate value can be 8 bits such as

```
CMP R1, #0xFF
```

### 8.5.2.2   CMN Compare Negate

The CMN has the following format:

```
CMN Operand1, Operand2
```

The instruction will add operand1 with operand 2 and set appropriate flag bit.

**Example 8.5**  Assume R1 contains 0x00000024 and R2 contains 0x13458978: the operation CMN R1, R2 with result carry and set C flag to 1.

### 8.5.2.3   TST (Test Instruction)

The test instruction has the following format:

```
TST Oprand1, Operand2
```

The test instruction performs AND operation between operand1 and Operand2 and sets appropriate flag bit. The operand can be immediate value or register such as

*TST R1, R2* ; This instruction performs R1 AND R2 operation and sets the appropriate flag.

OR

**TST R1, immediate,** the immediate value can be 8 bits such as
**TST R1, 0xFF**
*TEQ R1, R2* ; This instruction performs R1 Exclusive OR R2.

If R1 is equal to R2, then Z flag is set to 1.


### 8.5.3   Register Swap Instructions (MOV and MVN)

(a) **MOV Instructions:** The mov instruction has the following forms:
   MOV{S}{cond} Rd, Rn
      MOV{cond} Rd, #imm16
      The MOV instruction move the contents of register Rn to register Rd or moves 16 bits constant to Rd
      MOV R3, R2 or MOV R5, 0x34567
      The maximum size for immediate value (constant) is 16 bits or 4 digits in hex
      MOV R2, 0X3456789, this instruction results error
      The assembler offers pseudo instruction LDR Rn, = # constant to move a number more than 16 bits to a register such as LDR R1, = #0x34567890
(b) The MVN has the following Syntax
   MVN{S}{cond} Rd, Operand2

   The operand2 can be a register or 16 bit immediate data,
      THE MVN instruction complements (NOT) operand2 and move it to register Rd


**Example 8.6** What is the content of R1 after execution of the following instructions?
   Assume R2 contains 0XFFFF.

a. MOV R1, R2 ; R1 ← R2
   R2 = 0x0000FFFF
b. MVN R1, R2 ; R1 ← NOT R2
   R2 = 0xFFFF0000
A. MOV{S}{condition} Rd, immediate value

Immediate value is 16 bits, the range of immediate value if from 0x00000000 to 0x0000FFFF.

**Example 8.7** MOV R2, # 0x45 , the contents of R2 will be 0x00000045
**B. MOV Rn, Rm, lsl # n** ; shift Rm n times to the left and store the result Rn
**C. Conditional MOV**
**MOVEQ R2, 0x56** ; if zero bit is set then executes MOVEQ

## *8.5.4 Shift and Rotate Instructions*

ARM combined the rotate and shift operation with other instructions; the ARM processor performs the following shift operations.

> LSL   Logical Shift Left
> LSR   Logical Shift Right
> ASR   Arithmetic Shift Right
> ROR   Rotate Right

### 8.5.4.1   Logical Shift Left (LSL)

In logical shift left operations, each bit of register shifted to the left as shown in Fig. 8.5 and a zero will be placed in the least significant bit, the logical shift left multiplies the contents of register by two.

**LSL R1, R1, n** , shift to left R1 n times and store result in R1

**Example 8.8** What is the content of R1 after executing the following instruction? Assume R1 contain 0x00000500.

```
LSL R1, R1, 8
R1= 0x00050000
```

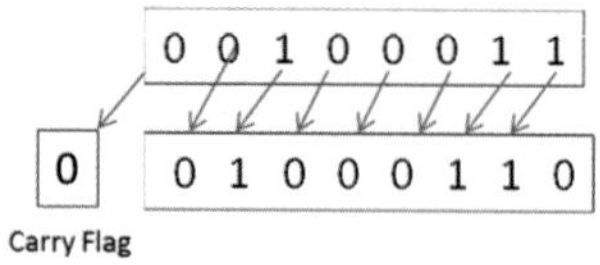

**Fig. 8.5** Logical shift left

### 8.5.4.2  Logical Shift Right (LSR)

In logical shift right operation, each bit of register shifted to the right as shown in Fig. 8.6, and a zero will be placed in the most significant bit; the logical right divides the contents of register by two.

```
LSR R1, R1, n ,shift to right R1 n times and store result in R1
```

**Example 8.9** What are the contents of R1 after executing the following instruction: assume R1 contains 0x00000500.

```
LSR R1, R1, 4
R1= 0x00000050
```

### 8.5.4.3  Arithmetic Shift Right (ASR)

In arithmetic shift right, the most significant bit does not change and each bit shifted to the right as shown in Fig. 8.7.

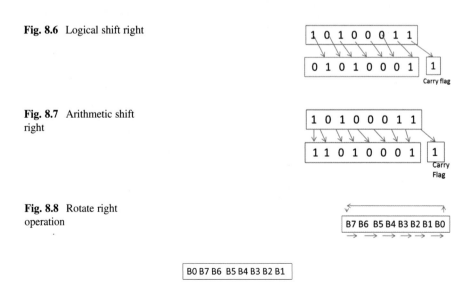

**Fig. 8.6** Logical shift right

**Fig. 8.7** Arithmetic shift right

**Fig. 8.8** Rotate right operation

**Fig. 8.9** One bit rotate right operation

**8.5.4.4  Rotate Right**

Figure 8.8 shows an 8-bit register and Fig. 8.9 shows the register after rotating one time.

**Example 8.10** What is the content of R1 after rotating 16 times? Assume R1 contains 0X0000FFFF

```
ROR R1, R1 , #16
R1= 0xFFFF0000
```

ARM combines data processing instructions and shift operation; shift operation is applied to the second operand of the instruction.

**Example 8.11** Register R2 contains 0XEEEEFFFF, by executing.

```
MOV R1, R2, ROR # 16 ;the R2 rotate 16 times and store results in R1
```

by rotating 16 times the contains of R1 will be xFFFFEEEE

*ADD R1, R2, R3, LSL #4* ;R1= R2 + R3 x $2^4$, R3 is shifted 4 times to the left and result is added to R3 and placed in R1.

Also a register can hold number of times the operand2 must be shifted.

| | |
|---|---|
| *ADD R1, R2, R3, LSL R4* | ;R1 = R2 + R3 X $2^{R4}$, Number of times R3 to be shifted is in R4. |
| *MOV R0, R1, LSL #3* | ;Shift R1 to the left three times and move the result to R0. |

## 8.5.5  ARM Unconditional Instructions and Conditional Instructions

Figure 8.10 shows the general format of an ARM instruction. ARM instruction defines two types of instructions, namely:

1. Unconditional instruction
2. Conditional instruction

**Fig. 8.10** General format of an ARM instruction

| Condition Code | Instruction |
|---|---|
| 31          28 | 27                                                    1 |

Condition code defines the type of instruction. If this field is set to 1110, then the instruction is an unconditional instruction, otherwise the instruction is a conditional instruction. To use an instruction as a conditional instruction, the condition will suffix to the instruction. The suffixes are as follows:

```
Condition Code Condition
  0000 EQ equal
  0001 NE not equal
  0010 CS carry set
  0111 CC carry is clear
  0100 MI negative (N flag is set)
  0101 PL positive (N flag is zero)
  0110 VS overflow set
  0111 VC overflow is clear
  1000 HI higher for unsigned number
  1001 LS less than for unsigned number
  1010 GT greater for signed number
  1011 LT signed less than
  1100 GT Greater Than
  1101 LE less than or equal
  1110 AL unconditional instructions
  1111 Unused code
```

The processor checks the condition flag in CPSR before executing the conditional instruction. If it matches with the condition of instruction, then the processor executes the instruction, otherwise skips the instruction.

*ADDEQ R1, R2, R3* ; If zero flag is set and it will execute this instruction.

**Example 8.10** Convert the following HLL to ARM assembly language.

```
If R1=R2 then
ADD R3, R4, R5
Endif
```

ARM assembly language for the above program would be:

```
CMP R1, R2
ADDEQ R3, R4, R5
```

**Example 8.11** Convert the following HLL to ARM assembly language.

```
If R1 = R2 Then R3= R4-R5
Else
If R1>R2 Then R3=R4+R5
```

ARM assembly language for the above program would be:

```
CMP R1, R2
SUBEQ R3, R4, R5
ADDGT R3, R4, R5
```

## 8.6   Stack Operation and Instructions

Part of the memory is used for temporary storage is called stack; the stack pointer (SP) holds the address of top of the stack as shown in Fig. 8.11. The stack is used as temporary memory, the program needs to save the contents of a register for use later on, then it will save the register contents in stack.

The register R13 is assigned as stack pointer (SP), and the stack uses the following instruction.

   **a. Push {condition} {Rn}:** transfer the contains of Rn into stack and subtract 4 from the stack pointer

**Example 8.12** Assume the content of R3 is 0x01234567; Fig. 8.12 shows the contents of stack and SP after executing push {R3}.

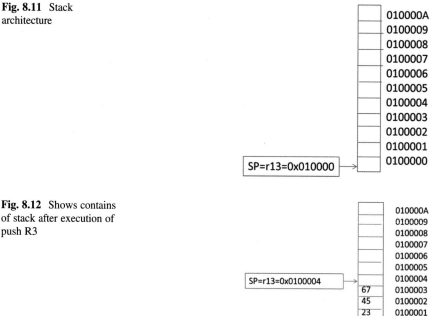

**Fig. 8.11** Stack architecture

**Fig. 8.12** Shows contains of stack after execution of push R3

**Fig. 8.13** Show stack after push operation

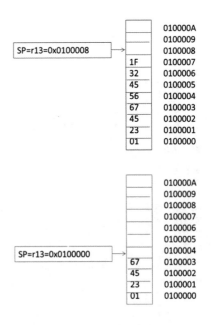

**Fig. 8.14** Contents of stack after POP operation

**Example 8.14** Figure 8.13 shows contents of stack and SP after execution of Push {R4}; assume R4 contains 0X5645321F.

**POP Instruction:** the POP instruction has the following format
`POP{condition} {Rn}`
`POP Rn`: the pop instruction remove the word from top of the stack and store it into register Rn and automatically increment stack pointer by 4

**Example 8.15** Figure 8.14 shows the content of stack and SP of after execution POP {R0}; the contents of R0 will be 0x5645321F and SP = 2000008.

## 8.7 Branch (B) and Branch with Link Instruction (BL)

The branch instruction has the following general format.

`B{condition} label`
`B label` ; branch to location label.
`BEQ label` ; if flag bit Z=1, then execute this instruction
`BL Subroutine` ; it will branch to subroutine and save contents of PC (R15) to R14 (link register) for return from subroutine.

For returning from subroutine to main program can use one of the following instructions
BX R14 or MOV R15, R14

**Example 8.16** Write a subroutine to find the value of Y = 16X + 4; assume R1 holds the Y and R2 holds X.

```
BL Funct
Funct SUB R1, R1, R1
ADD R1, R1, R2, LSL4
ADD R1, R!, #04
 BX  R14 ; Move return address to PC
```

**Example 8.17** Rewrite following assembly language using conditional instructions.

```
CMP R1,R2
BEQ Exit
ADD R1, R2, R3
Exit:
SUB R1, R5, R6
```

By using conditional instructions, the above assembly language can be represented by

```
CMP R1,R2
SUBEQ R1,R5,R6
ADDNE R1,R2,R3
```

## 8.8   Multiply (MUL) and Multiply-Accumulate (MLA) Instructions

```
The Multy instructions use following instruction
MUL{cond} Rd, Rm,Rs ;Rd= Rm*Rs
MLA Multiply and Accumulate
MLA(cond} Rd,Rm,Rs, Rn ; Rd= Rm*Rs +Rn
```

## 8.9   Summary

- The function of compiler is to convert the HLL to the assembly language.
- The function of assembler is to convert the assembly language to the machine code (binary).
- The computer instruction is represented by mnemonic form such as "ADD."
- Each instruction may have one or two or three operands, ADD R1, R2, and R3, where R1, R2, and R3 are called operands.
- ARM stand for Advanced RISC Machine and ARMv7 uses 32-bit and 16-bit instruction.

- ARMv7 contain 31 registers and only 16 registers R0 through R15 used by programmer.
- The Register R15 is used for the program counter (PC), R14 is used for the link register (LR), and R13 is used for the stack pointer (SP).
- The PSR register is used to store control bits (I, F, M, and T) and flag bits (N, Z, C, and V).
- The ARM processor offers two types of instructions, and they are unconditional and conditional instruction.
- The instructions CMP, CMN, TST, and TEQ will set processor status register.
- Part of the memory is used for the stack, and the stack pointer holds the address of the top of the stack.

Chapter 9 covers how to use Keil development tools in order to run assembly language, how to use μVision debugger, Programming Rules, **data representation and memory, and directives**

## Problems and Questions

1. Explain how HLL converted to machine code.
2. List types of instructions based on number of operands.
3. Which register of ARM processor is used for the program counter (PC)?
4. Which register of ARM processor is used for stack pointer (SP)?
5. Which register of ARM processor is used for link register?
6. What is contents of R5 after execution of following instruction, assume R2 contains 0X34560701 and R3 contains 0X56745670

   (a) ADD R5, R2, R3
   (b) AND R5, R3, R2
   (c) EOR R5, R2, R3
   (d) ADD R5, R3, #0x45

7. Trace the following instructions

   MOV R1, 0x25 R1=
   ADD R2, R1, # 0x97 R2=

8. Trace the following instructions

   MOV R1, #0x10
   MOV R2, #0x20
   MOV R3, 0x0F
   CMP R1, R2
   ADDGT R3, R1, R2 R3=
   SUBLE R4, R2, R1 R4=

9. Trace the following instructions

   MOV R1, #0x0F
   MOV R2, #0x23
   AND R4, R2, R1  R4=

10. What is contents of R3?

    MOV R1, #0x52
    LSL R3, R1, #0x8  R3=

11. What are the contents of R1? Assume R2 = 0x00001234.

    (a) MOV R1, R2, LSL #4
    (b) MOV R1, R2, LSR #4

12. What is the difference between these two instructions?

    SUBS R1, R2, R2
    SUB R1,R2, R2

13. Convert the following HLL language to ARM instructions.

```
IF R1>R2 AND R3>R4 then
  R1= R1 +1
  Else
  R3=R3 +R3*8
  Endif
```

14. What is contents of R1 after executing following Instruction assume R1=0x11245600

    LSR R1, R1 , #8  R1=

15. What is contents of R1 after executing following Instruction assume R1=0xF1245678

    ROR R1, R1, #8  R1=

16. Convert the following HLL language to ARM instructions.

```
IF R1>R2 OR R3>R4 then
  R1= R1 +1
  Else
  R3=R3 +R5*8
  Endif
```

17. Convert the following flowchart to ARM assembly language.

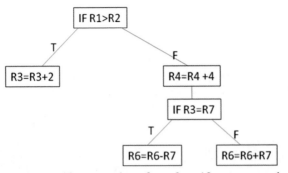

18. Write a program to add ten numbers from 0 to 10 or convert the following C language to ARM assembly language.

```
int sum;
   int i;
   sum = 0;
   for (i = 10 ; i > 0 ; i - - ){
   sum = sum +1
   }
```

19. Write a program to convert the following HLL to ARM assembly.

```
a= 10;
  b=45;
  while ( a! =b ) {
  if (a < b)
  a = a +5;
  else
  b= b+5;
  }
  SOLUTION:
```

20. Convert the following HLL to ARM assembly.

```
IF R1>R2 AND R3 >R4 then
   R1= R1 +1
   Else
   R3=R3 +R5*8
   Endif
```

21. Convert the following flowchart to ARM assembly.

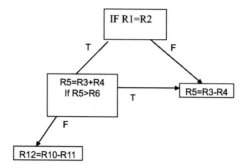

# Chapter 9
# ARM Assembly Language Programming Using Keil Development Tools

**Objectives: After Completing this Chapter, you Should Be Able to**
- Explain the function development tool.
- Explain the function of cross-assembler.
- List several development tools for running assembly language program.
- Install the Keil development tools.
- Run and debug a program.
- Use program template to write you own program.
- Learn programming rules.
- Represent data in memory for an assembly language program.
- Learn the application of directives.
- Distinguish the different types of data directives.
- Run a step-by-step program and observe the contents of each register.

## 9.1 Introduction

Processor manufacturers publish documentation that contains information about their processors, such as lists of registers, the function of each register, size of the data bus, size of the address bus, and a list of the instructions that can be executed. Each CPU has a known instruction set that a programmer can use to write assembly language programs. Instruction sets are specific to each type of processor. For example, Pentium processors implement a different instruction set than ARM processors. Programs written using the instruction set of a processor are said to be written in assembly language. The function of an **assembler** is to convert assembly language to machine code (binary) that the CPU can execute.

When an assembler runs on one processor but can assemble instructions for a different processor with a different instruction set it is called a **cross-assembler**. Processor simulators are a key development tool, since they allow for a controllable test environment in a setting such as a Windows or Linux workstation. They may

© The Author(s), under exclusive license to Springer Nature Switzerland AG 2022
A. Elahi, *Computer Systems*, https://doi.org/10.1007/978-3-030-93449-1_9

also facilitate the transfer or download of the program to the target processor. The following development tools are some which support ARM processors:

1. ARM Keil Microcontroller Tools (www.keil.com)
2. IAR Embedded Workbench (www.iar.com)
3. GNU ARM Assembler (www.gnu.org)

## 9.2 Keil Development Tools for ARM Assembly

For the examples in this book, **Keil μVision® IDE** (integrated development environment) from Keil's **Microcontroller Development Kit (MDK) version 5** is used. A free version of this software can assemble and simulate the execution of ARMv7 instructions provided that the size is under 32 K. The download is available from Keil's website: http://www.keil.com.

When first installed, a dialog titled **Pack Installer** may open after the installer has finished. This utility assists the user in downloading and installing environments for μVision to enable the simulation of different boards and devices. By default, however, several device templates come pre-installed with μVision (Fig. 9.1).

- To get started, open μVision and select *Project → New μVision Project. . . .*
- Name your project and choose a location to save it in.

    After saving, a dialog will open and prompt you to *Select Device for Target "Target 1". . . .* Depending on whether or not you have installed any additional packs from the *Pack Installer*, this screen may look different. Several **ARM** processors are included with the default installation. For the examples in this book, the **ARM Cortex M3 (ARMCM3)** was selected.

- Select **ARM Cortex M3 → ARMCM3** and press OK. (Fig. 9.2).

For each processor, μVision has several libraries available. Some are essential, such as start-up configuration, while others are optional extensions to enable broader functionality, such as Ethernet drivers and Graphics interfaces. To run simple ARMv7 assembly programs, only the CORE and Start-up component are needed.

- Select **CMSIS → select CORE**
- Select **Device → Startup** and select DEPRECETED and hit OK. (Fig. 9.3)

    Now, you should have a blank project openin the *Project* pane. Next, the project needs to be configured to use the simulator to run the programs.

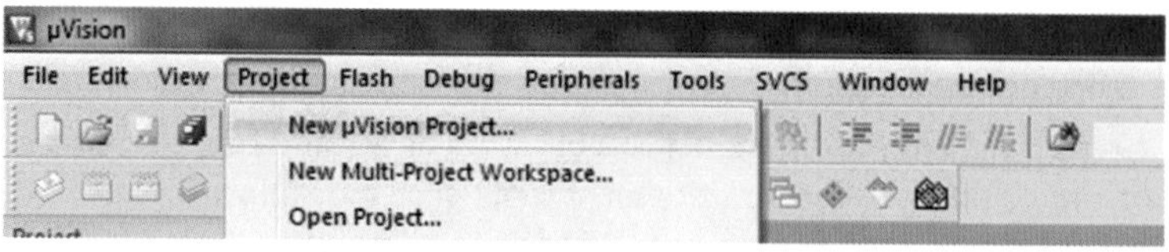

**Fig. 9.1** Creating a new project in Keil μVision® IDE v5.22

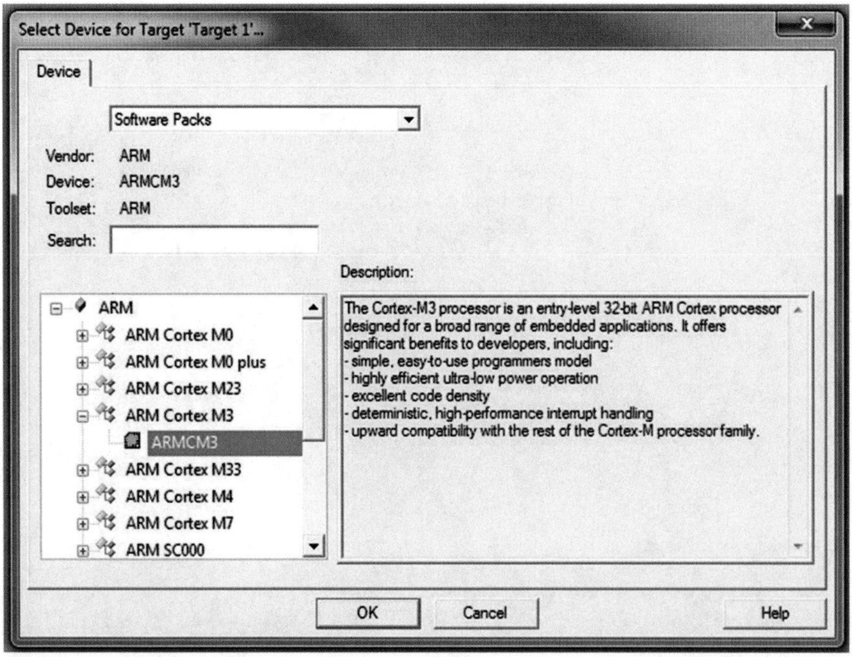

**Fig. 9.2** Selecting the ARM Cortex M3 processor

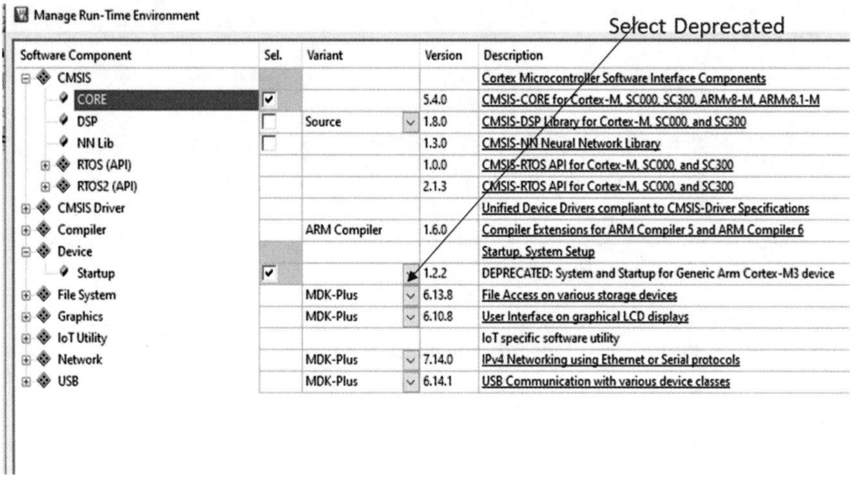

**Fig. 9.3** The Manage Run-Time Environment setup

- Right click on the *Target 1* folder and select *Options for Target "Target 1"*…
  (Fig. 9.4).
- Click on the *Debug* tab and select *Use Simulator.* (Fig. 9.5)

**Fig. 9.4** Changing the project configuration

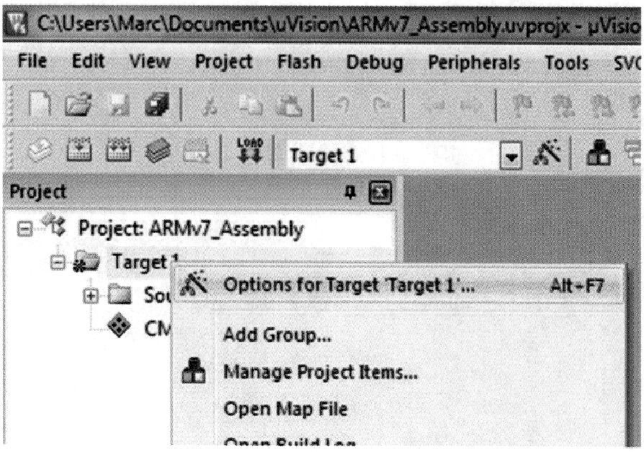

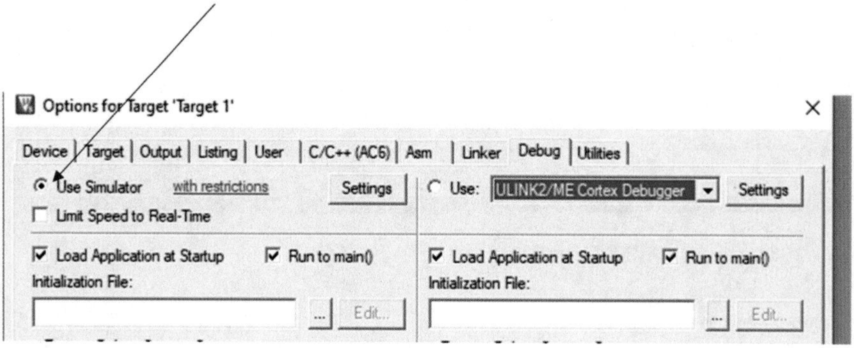

**Fig. 9.5** Setting the project to use the simulator

- Click on Target and select Use MicroLIB then click ok as shown in Fig. 9.6
- Click on Target1 then Device. If there is a red "X" on Device in the project window, this means that the target must be updated, right click on startup_ARMCM3.s and startup_ARMCM3.c one at a time and select update config file as shown in Fig. 9.7.

On the µVision window select File then New to open a blank workspace to start typing your program. Save the program with extension ".s" for an assembly program, or with the extension ".c" for a C language program. This example will add two numbers as shown in Fig. 9.8.

Add your program to the Source Group 1 by right clicking on Source Group1 and selecting add existing file to Group. Find the file, click add then close the file explorer as shown in Fig. 9.9.

To verify the above action, click on the "+" to the left of Source Group1, and it will display the file name as shown in Fig. 9.10.

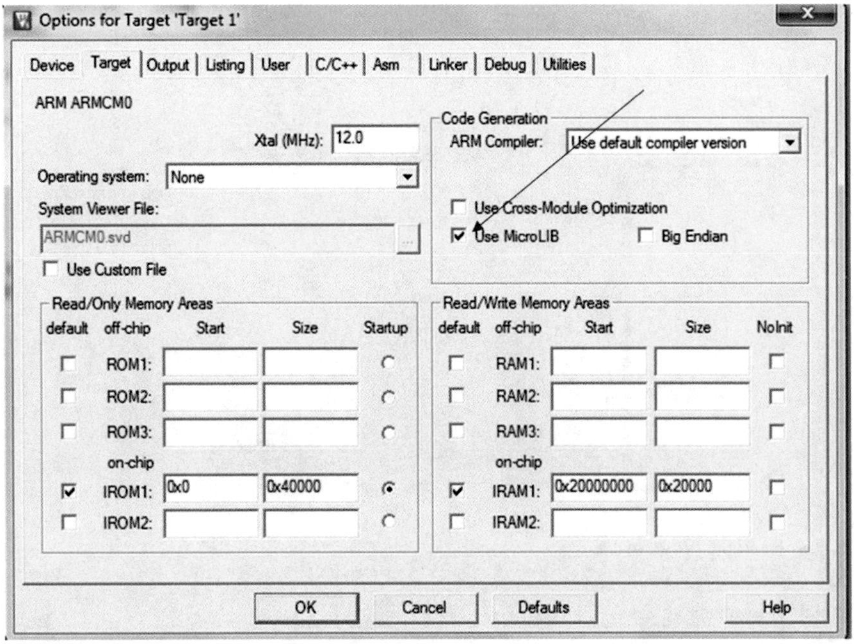

**Fig. 9.6**   Options for Target1 (Use Micro LIB)

**Fig. 9.7**   Updated target device

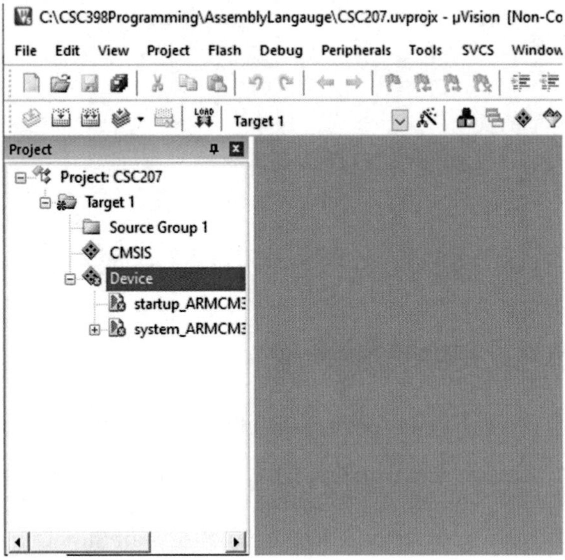

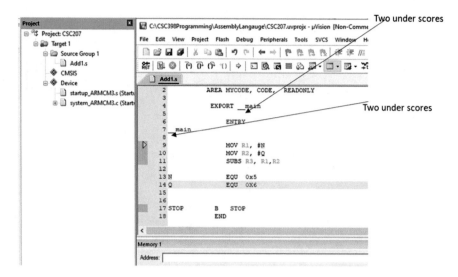

**Fig. 9.8** Example program file

**Fig. 9.9** Adding file to the group

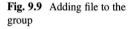

## 9.2.1 Assembling a Program

Once the source code is ready to be assembled, perform the following steps:

1. Click on Translate or **Ctrl + F7** to build current active file and check for error
2. Click on build or F7 build all target files and check for error as shown in Fig. 9.11
3. Running program, select debug then Start/Stop Debug then Run (F5)

The **Build Output** panel on the bottom of the window will show any errors, warnings, or if the project was built successfully. A successful build should look like Fig. 9.12, whereas a failure will give error descriptions to help the programmer find where the code is incorrect.

**Fig. 9.10** Verifying the file added to the group

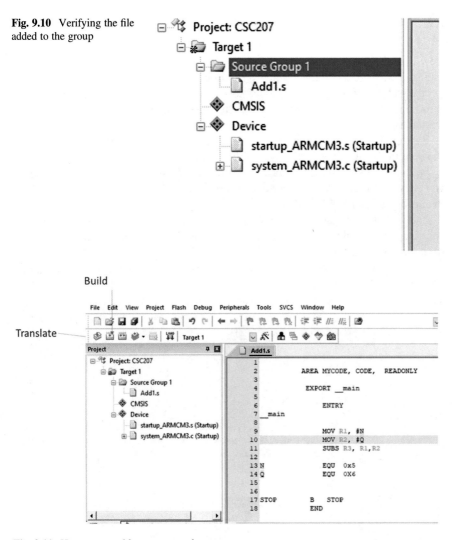

**Fig. 9.11** How to assemble program and run a program

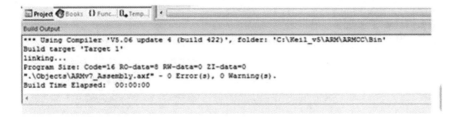

**Fig. 9.12** Build Output for a successful assembly

## 9.2.2 Running the Debugger/Simulator

Now that you have compiled a piece of code, you will want to debug the code for testing. To start the debugger, click on **Debug → Start/Stop Debug Session** from the menu bar. Once in debug mode, keep clicking on the Step One Line button shown in Fig. 9.13 to step through the startup code and get to the assembly program written earlier.

By stepping through an executing the startup code, the example program will show up in the main μVision window like in Fig. 9.14.

By default, the registers will be displayed in the *Registers* pane on the left side. If you click the "+" next to the *xPSR* line, you can view the current *flag bits*—this is the **Program Status Register (PSR)**.

The *Disassembly* pane will display some of the current data, such as the machine code instructions and notes of what line in the source code is being executed, and at what memory address.

To view the current values in memory while debugging, open the Memory 1 panel in the bottom right corner (Fig. 9.15). Up to four windows are available, and by default *Memory 1* should be enabled. If it is not, it may be enabled under the *Menu* bar by clicking **View → Memory Windows → Memory 1**. *(An enabled window will have this icon next to it:▦).*

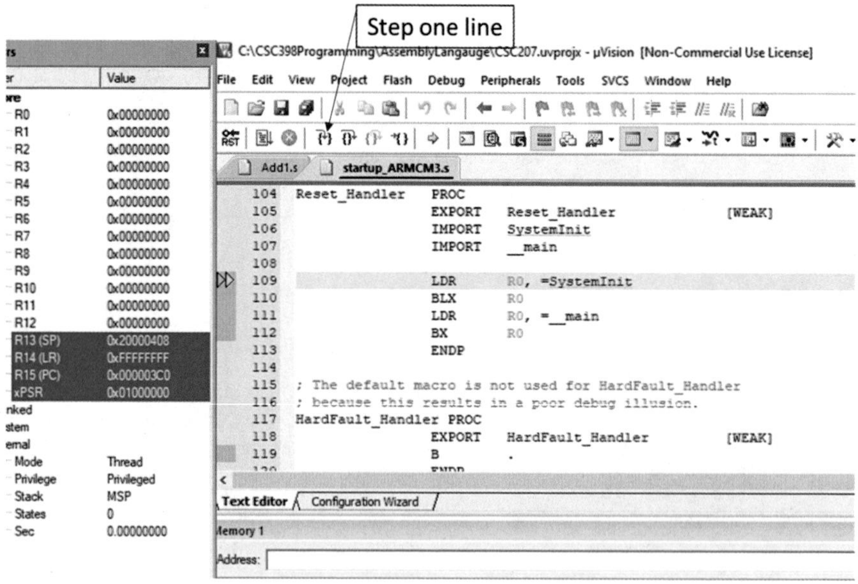

**Fig. 9.13** Debugger shows registers in left pan and startup instructions in right pan

To search memory, enter the hexadecimal address of the portion of memory you would like to view. Since it is a byte-addressable processor, each address holds one byte of information, and is displayed as two hexadecimal digits.

## 9.3 Program Template

Figure 9.16 contains a sample template of code to write a program for the ARM Cortex M3. The source code located between the __main and *STOP* label is executed when the program is run. *(All labels must have no whitespace like spaces or tabs on the left—they must be the first text in the column, and the first characters on the line.)*

## 9.4 Programming Rules

### 9.4.1 CASE Rules

Instructions, symbols, and labels can be written in uppercase or lowercase but cannot be combined (e.g., MOV or mov is correct, but MoV or moV is not).

### 9.4.2 Comments

The programmer can write comments after a semicolon (;)

```
MOV R1, R2; Moving contents of R2 to R1
```

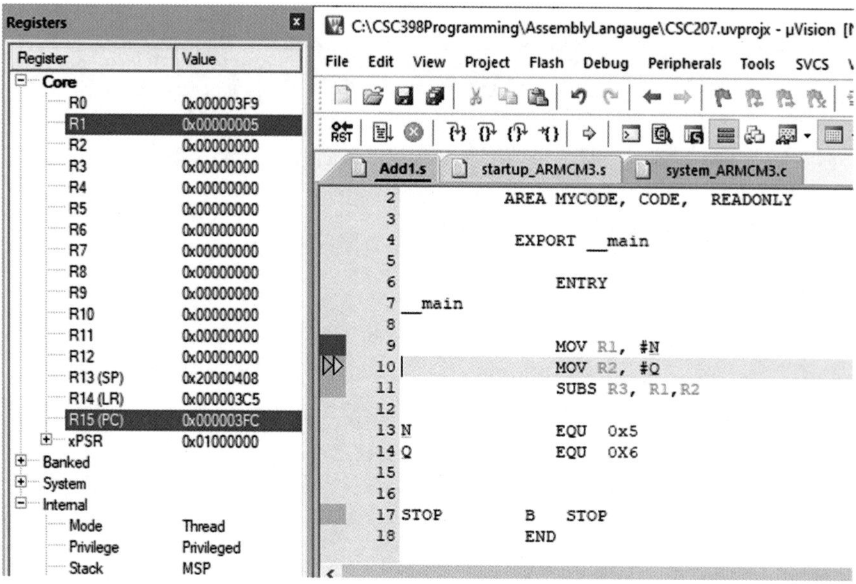

**Fig. 9.14** Shows registers on the lefy pan and program in the right pan

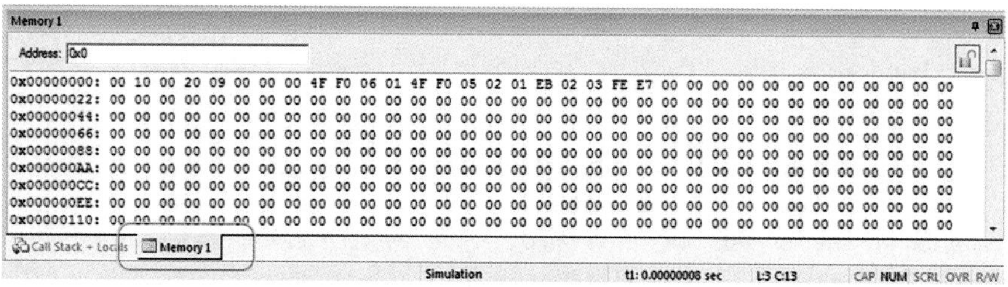

**Fig. 9.15** Location of the memory window and panel tab

## 9.5   Data Representation and Memory

ARM processors define a word as 4 bytes and a half word as 2 bytes. Data can be represented in the form of hexadecimal, decimal, and binary numbers (Fig. 9.17).

Data and code are held in memory. Figure 9.18 shows a block diagram of memory. The *address* defines the location of the data in memory. Each location of an ARM processor's memory holds one byte. In assembly language, a label—as shown in Fig. 9.18—represents the address in memory for the data. When the code is translated, the assembler automatically decides what address to use in place of the label and substitutes it where appropriate in the program.

| | | |
|---|---|---|
| **Fig. 9.16**  Program template for ARMCM3 processor in Keil μVision v5 | AREA  MYCODE, CODE,  READONLY | |

EXPORT __main

ENTRY

__main

**PROGRAM  CODE**

STOP          B  STOP

END

| Base | Prefix | Example |
|---|---|---|
| Decimal (10) | - | 17 |
| Hexadecimal (16) | 0x | 0x11 |
| Binary (2) | 2_ | 2_00010001 |

**Fig. 9.17**  Numerical representation formats

| Label | Data |
|---|---|
| List | 0x23 |
| List+1 | 0x4A |
| List+2 | 0x56 |
| List+3 | 0xF5 |

| Memory Address | Data |
|---|---|
| 0x400 | 0x23 |
| 0x401 | 0x4A |
| 0x402 | 0x56 |
| 0x403 | 0xF5 |

**Fig. 9.18**  Block diagram of memory if *List* is stored at address 0x400

Since each memory location holds one byte, the programmer must be careful when working with *words* and *half words*. Processors can function with either **Big Endian** or **Little Endian** ordering. Figure 9.19 shows the number 0x2000000F stored at 0x400 in both systems.

- The **ARM Cortex M3** used for these examples is a **Little Endian processor**.

| | Big Endian | Little Endian |
|---|---|---|
| 0x400 | 0x20 | 0x0F |
| 0x401 | 0x00 | 0x00 |
| 0x402 | 0x00 | 0x00 |
| 0x403 | 0x0F | 0x20 |

**Fig. 9.19** Big vs. Little Endian

## 9.6 Directives

A directive is an assembler command that is executed by the assembler. Directives never produce any machine code. Directives are used to indicate the start of code or data and the end of the program. A simple directive is END, which marks the end of a program. Some of the most useful directives used by the ARM Assembler are as follows:

- **AREA**—defines a segment of memory
- **ENTRY**—defines the start of the program
- **EQU**—used to assign a constant to a label

```
X EQU 0x6        ;x is label
MOV R1, #X ; R1 =0x6
```

- **ALIGN- Align the data or code in memory locations**
- **PROC- Define the start of procedure**
- **ENDP-define the end of procedure**
- **END- Enad of the program**

### 9.6.1 Data Directive

Data directives define the types and size of data.

- **DCB** (Define Constant Byte)
- **DCW** (Define Constant Half Word)
- **DCD** (Define Constant Word)
- **SPACE** (Reserve a zeroed block of memory)

### 9.6.1.1 DCB (Define Constant Byte)

This directive is used for allocating one or more bytes in memory.

```
list1 DCB 0xF,10,2_00010001
    list1 : 0x0F
    list1 + 1: 0x0A
    list1 + 2: 0x11
```

### 9.6.1.2 DCW (Define Constant Half Word)

This directive defines constant half words (16 bits, 2 bytes) and requires two memory locations per half word.

```
list1 DCW 0xFF00, 0x13
    list1   : 0x00
    list1 + 1: 0xFF
    list1 + 2: 0x13
    list1 + 3: 0x00
```

### 9.6.1.3 DCD (Define Constant Word)

DCD is used to define a word (32 bits, 4 bytes) and requires four memory locations per word.

```
list1 DCD 0x12345678, 0xFF
    list1   : 0x78
    list1 + 1: 0x56
    list1 + 2: 0x34
    list1 + 3: 0x12
    list1 + 4: 0xFF
    list1 + 5: 0x00
    list1 + 6: 0x00
    list1 + 7: 0x00
```

### 9.6.1.4 Character Strings

A sequence of characters is called a character string. In ARM assembly, strings must be *null-terminated* in that they must end with a 0 value when they are defined.

```
List1 DCB "Assembly",0
List2 DCB "I have $250",0
```

The assembler breaks the string into bytes and stores them each in Little Endian order.

### 9.6.1.5   Single Character

When storing a single character in a register or memory location, the character must be inside single quotation marks. The assembler converts the ASCII to hexadecimal.

```
List DCB 'A'
```

### 9.6.1.6   SPACE

Reserves memory locations for later use.

```
List SPACE 20
```

Reserves 20 bytes of memory starting at *List*.

## 9.7   Memory in μVision v5

Memory in μVision is simulated, and the program tries to replicate real-world conditions. Many microprocessors use a combination of ROM (read-only memory) and RAM and will have a program stored in ROM to respond to a power-up, reboot, or other situation where RAM will not be propagated or reliable.

To do this, memory is marked as none, some, or all of the following permission flags:

- *Exec*—Memory holds instructions and can be executed.
- *Read*—Memory can be read.
- *Write*—Memory can be written to.

By default, the area in memory that most directives will store their data in is marked as both *exec* and *read*, but not *write*. This mimics ROM since the instructions stored there can be read and run, but attempts to write to those locations (with STR or similar) will not work. The assembler is also unlikely to notify you of this result.

**Example 9.1**  What are the contents of *list1* after execution, if stored in *ROM*?

```
ADRL R0, list1 ;store the address of list1 in R0
MOV R1, #2 ;store 2 in R1
LDRB R2, [R0] ;load the byte at list1 into R2
ADD R3, R1, R2 ; R3←R1+R2 (0+2 = 2)
STRB R3, [R0] ;store R3 (#2) at list1
List1 = 0x00
```

While µVision and ARM source libraries allow for programs to more easily utilize data stored in RAM rather than ROM, it goes beyond the scope of this chapter. However, knowing what addresses are marked as *read* and *write*, you can still use RAM to store data into memory.

By default, for the **ARMCM3** processor, memory locations 0x20000000 – 0x20020000 are marked as *read* and *write* RAM in µVision.

**Example 9.2** What are the contents of memory at 0x200000000 after the execution of these instructions?

```
list1 DCB 0x0
ADR R0, list1 ; store the address of list1 in R0
MOV R1, #2 ;store 2 in R1
LDRB R2, [R0] ;load the byte at list1 into R2
ADD R3, R1, R2 ; R3←R1+R2 (0+2 = 2)
MOV R4, #0x20000000 ; store our initial RAM address in R4
STRB R3, [R4] ;store R3 (#2) at 0x20000000
List1 = 0x00
0x20000000 0x02
```

## 9.8  Summary

- The simulation of a processor which runs on different processor is called cross-assembler.
- There are several development tools for the ARM processor such Keil, IAR, and GNU.
- The ARM instructions and labels can be written in uppercase or lower case but cannot combined.
- The ARM processor defines word as 4 bytes, half word as two bytes, and they can be represented in binary (2_1000011), hex (0x24), or decimal 45.
- The data directives are DCB (Define Constant Byte), DCW (Define Constant Half Word), DCD (Define Constant Word), and SPACE (reserved a zeroed block of memory).
- The DCB is used to represent one byte in memory location such as

  LIST DCB 0x32

- The DCW is used to represent half word in memory location such as

  LIST DCW 0x3245

- The DCD is used to represent a word in memory location such as

  LIST DCD 0x87673245

- The SPACE directive is used to reserve memory location with zero values in all of the locations such as

LIST SPACE 20

- The character string must be terminated by null (0) such as

List DCB "WELCOME," 0

- The single catheter in a memory location must be inside of single quotations such as

List DB "A"

- Chapter 10 covers load, store, pseudo instructions, bits field instructions, and ARM addressing modes

## Questions and Problems

1. List Data Directives
2. Show how the following data are stored in memory using Little Endian

   List DCB 0x34, 0x22, 0x67,0x56

3. Show how the following data are stored in memory using Little Endian

   List DCW 0x534, 0x22, 0x167,0x5692

4. What is application of ADR instruction
5. Show how the following data are stored in memory using Little Endian

   List DCD 0x53456721, 0x00922, 0x16789677, 0x569234.

6. What is application of ADR instruction?
7. Represent CSC207 Computer system in the form of character string for ARM assembly
8. Why character string terminated by null character in ARM assembly

# Chapter 10
# ARM Instructions Part II and Instruction Formats

**Objectives: After Completing this Chapter, you Should Be Able to**
- Explain the different types of load instructions.
- List the different types of store instructions.
- Distinguish the different types of ARM addressing mode.
- List the ARMv7 pseudo instructions.
- Learn the application of ADR and LDR instructions.
- Explain the bit field instruction operation.
- Learn how the data are represented in memory.
- Able to convert data processing instructions to machine code.
- Able to convert LDR, STR, LDRB, and LDRH to machine2 code.

## 10.1    Introduction

The data transfer instructions are used to transfer data from memory to registers and from registers to memory. ARM processor used LDR and STR instructions to access memory. LDR and STR able to use register indirect, pre-index addressing, and post-index addressing to access memory. ARM offers several pseudo instructions which are used by programmer and assembler to convert them to ARM instructions.

## 10.2    ARM Data Transfer Instructions

**Load Instructions (LDR)**
   The LDR instruction is used to read data from memory and store it into a register, and it has the following general format.

   **LDR[type]{condition} Rd, Address**
      Where "type" defines the following load instructions

© The Author(s), under exclusive license to Springer Nature Switzerland AG 2022          213
A. Elahi, *Computer Systems*, https://doi.org/10.1007/978-3-030-93449-1_10

LDR     Load 32 bits (word)
LDRB    Load 1 byte
LDRH    Load 16 bits (half word)
LDRS    Load signed byte
LDRSB   Load sign extension
LDRSH   Load signed half word
LDM     Load multiple words

Condition is an optional such as LDREQ load data if Z flag $=1$ and Rd is destination register.

**Example 10.1** Assume R0 holds address 0000 and the following memory is given, show the contents of R1 and R3 after executing the following instructions.

| Address | Contents |
|---------|----------|
| 0 | 0X85 |
| 1 | 0XF2 |
| 2 | 0X86 |
| 3 | 0XB6 |

LDRH R1, [R0]   R1 = 0x0000F285
LDRSH R3, [R0]   R3 = 0xFFFFF285

## 10.2.1  ARM Pseudo Instructions

ARM supports multiple pseudo instructions; the pseudo instruction is used by the programmer, and assembler converts the pseudo instruction to ARM instruction.

**ADRL Pseudo Instruction**
    ADR is used to load the address of memory location into a register and has the following format.

**ADRL Rd, AddressExample 10.2** The following instructions will read the address of data and then load the data into register R3:

ADR R0, table   Move address represented by table
LDR R3, [R0]    R3 = 0x23456780
Address         Data
Table           0x23456780

**LDR Pseudo Instruction**

LDR pseudo instruction is used for loading a constant into a register. In order to move a 32 bits contestant into a register, the instruction MOV Rd, #value only can move 12 bits to the register Rd because the operand2 in instruction format for MOV is 12 bits. The LDR pseudo instruction has the following format.

**LDR Rd, = ValueExample 10.3** The following instruction will load the R1 with 0x23456789:

LDR R1, =0x23456789

## 10.2.2  Store Instructions (STR)

The STR instruction is used to transfer contents of a register into memory and has the following general format.

**STR[type]{condition} Rd, [address]**
Where "type" defines the following instruction types

| | |
|---|---|
| STR | Store 32 bits (word) |
| STRB | Store 1 byte |
| STRH | Store 16 bits (half word) |
| STM | Store multiple words |

**Example 10.4**

| | |
|---|---|
| **STR R5,** | Store contents of R5 into the memory location that R3 holds the |
| **[R3]** | address; R3 is the base register |

## 10.3  ARM Addressing Mode

- The ARM processor support ARM offers several addressing modes and they are pre-indexed, pre-indexed with immediate offset, pre-indexed with register offset, pre-index with scaled register, pre-index with register offset and write back, post-index with immediate offset, post-index with register offset, and post-index with scaled register offset; the following table shows a summary of ARM addressing modes.

| Addressing mode | Assembler syntax | Effective address (EA) |
|---|---|---|
| Immediate | MOV R1, #0X25 | Data is part of instruction |
| Pre-indexed | [Rn] | EA = Rn |

| | | |
|---|---|---|
| Pre-indexed with immediate offset | [Rn, # offset] | EA = Rn + offset |
| Pre-indexed with register offset | [Rn, ±Rm] | EA = Rn ± Rm |
| Per-index with scaled register | [Rn, Rm, Shifted] | EA = RN + Rm shifted |
| Pre-indexed with an immediate offset and write back | [Rn, offset]! | EA = Rn + offset Rn = Rn + offset |
| Pre-index with register offset and write back | [Rn, ±Rm,]! | EA = Rn ± Rm Rn = Rn ± Rm |
| Pre-index with scaled register offset and write back | [Rn, Rm, Shifted]! | EA = Rn ± Rm shifted Rn = Rn ± Rm shifted |
| Post-index with immediate offset | [Rn], offset | EA = Rn Rn = Rn + offset |
| Post-index with register offset | [Rn], ±Rm | EA = Rn Rn = Rn ± Rm |
| Post-index with scaled register offset | [Rn], ±Rm SHL #n | EA = Rn Rn = Rn ± Rm shifted |

## 10.3.1   Immediate Addressing

In immediate addressing, the operand is part of instruction such as

MOV R0, # 0x34
or
ADD R1, R2, #0x12

## 10.3.2   Pre-indexed

In pre-index addressing mode represented by [Rn], the effective address (EA) is contents of Rn such as

LDR R2, [R3]

### Pre-indexed with Immediate Offset
Pre-indexed with immediate offset is represented by [Rn, #offset] such as

LDR R0, [Rn, #Offset]
The offset is an immediate value such as

$$\text{LDR R1, } [Rn, \#0x25]$$
$$EA = R2 + 0x25$$

**Pre-indexed with Register Offset**
The offset can be register or register with shift operation:

$$\text{LDR R0, } [R1, R2]$$
$$EA = R1 + R2$$

**Example 10.5** What is the effective address of the following address? Assume R5 contains 0X00002345

[R5, #0x25]
$EA = 0x000002345 + 0X25 = 0x0000236A$

**Example 10.6** What is effective address of the following pre-index addressing, assume R5 = 0x00001542  and R2 = 0X00001000

[R5, R2]
$EA = R5 + R2 = 0X00001542 + 0X00001000 = 0X00002542$

**Pre-indexed with Scaled Register**
The offset contains register with shift operation:

LDR R0, [Rn, R2, LSL#2]

**Example 10.7** What is EA of the following instruction?

LDR R0, [Rn, R2, LSL#2]
$EA = Rn + R2 * 4$
R2 shifted to the left twice (multiply by 4) and added to Rn.

## 10.3.3   Pre-indexed with Write Back

The general format for pre-index addressing with write back is

[Rn, Offset]!
The exclamation (!) character is used for write back; the offset can be immediate
   value or register or shifted register:
$EA = Rn + \text{offset}$   and   $Rn = Rn + \text{offset}$

**Pre-index with Immediate Offset and Write Back**
LDR R0, [R1,# 4]! ; Exclamation mark mean update the register

$EA = R1 + 4$  and  R1 updated by $R1 = R1 + 4$.

**Example 10.8**  What is the effective address and final value of R5 for the following instruction? Assume the contents of R5 = 0x 00002456:

$$\text{LDR R0, [R5, \#0X4]!}$$
$$\text{EA} = \text{R5} + 0\text{x}4 = 0\text{x}000245\text{A}$$
$$\text{R5} = \text{R5} + 0\text{x}4 = 0\text{x}000245\text{A}$$

**Pre-index with Register Offset and Write Back**

$$\text{LDR R0, [R1, R2]!}$$
$$\text{EA} = \text{R1} + \text{R2} \quad \text{R1} = \text{R1} + \text{R2}$$

**Correction EA = R1 + R2, R1 + R2.**

**Example 10.9**  What is the effective address and final value of R5 of the following instruction? Assume the contents of R5 = 0x 00002456 and R2 0X00002222:

LDR R0, [R5, R2]!

$$\text{EA} = \text{R5} + \text{R2} = 0\text{x}00004678$$
$$\text{R5} = \text{R5} + \text{R2} = 0\text{x}00004678$$

**Pre-index with Scaled Register Offset and Write Back**
LDR R1, [Rn, R2, LSL#2]!

$$\text{EA} = \text{Rn} + \text{R2}^*4$$
$$\text{Rn} = \text{Rn} + +\text{R2}^*4$$

### 10.3.4   Post-index Addressing

The general format of post-index addressing is

LDR R0, [Rn], offset
Offset can be immediate value or register or shifted register.

**Post-index with an Immediate Value**
LDR R0, [Rn], #4

Effective address = Rn  and  Rn = Rn + 4.

**Post-index with Register Offset**

LDR R0, [Rn], Rm

Effective address = Rn and Rn = Rn + Rm

**Post-index with Scaled Register Offset**

LDR R0, [Rn], Rm, SHL#4

Effective address = Rn and Rn = Rn + Rm*16

## 10.4  Swap Memory and Register (SWAP)

The swap instruction combines the load and stores instructions into one instruction, and it has the following format.

**SWP Rd, Rm, [Rn]**

The register Rd is destination register, Rm Swap memory and register (SWAP) is the source register, and Rn is base register.

The swap instruction performs the following functions.

Rd ← memory [Rn] Load Rd from memory location [Rn]

[Rn] ← Rm store the contents of Rm in memory location [Rd]

*SWPB Rd, Rm, [Rn]*    Swap one byte

## 10.5  Storing Data Using Keil µVision 5

By default memory location 0x20000000 through 0x20020000 is reserved for writing and reading, for storing data at memory location list the address of list must be added to 0x20000000.

Example: write a program to store 5 at memory location list.

```
AREA MYCODE, CODE, READONLY
  EXPORT __main
  ENTRY
 __main

      MOV R1, #0x5
      ADRL R0, list
      LDR R3, = 0x20000000
          ADD R4, R0, R3
```

```
           STR R1, [R4]
           SUB r3, r1, r1

list DCB 0

STOP   B   STOP
           END
```

By using debugger results the R0 contains R0 is 0x421 and R4 is 0x2000434, by check memory location addressed by R4 should contains 0x5.

## 10.6   Bits Field Instructions

ARM offers two bit field instructions and they are bit field clear (BFC) and bit field insertion (BFI).

**BFC (Bit Field Clear Instruction):** BFC has the following general format.

**BFC {cond} Rd, # lsb, #width**
**Rd** is destination register.
**lsb** determines start of bit position in the source register (Rd) to be clear.
**Width** determines number of bits to be clear from lsb to msb of the Rd register.

**Example 10.10** Write an instruction to clear bits 7 through 15 of register R4; assume R4 contains 0xFFFEFEFE.

BFC R4, #7, #8 clear bit 7 through bit 15 (8 bits) of register R4.
The initial value in R4 is.

b31                            b15        b7        b0

| 1111 1111 1111 1110 1111 1110 1111 111  0 |

After clearing bit 7 through 15 of R4 results.

b31                            b15        b7        b0

| 1111 1111 1111 1110 10000000011 111  0 |

**BFI (Bit Insertion Instruction):** Bit insertion is used to copy a set of bit from one register Rn into register Rd starting from lsb of Rd; BFI has the following format.

**BFI{*cond*}Rd,Rn, #lsb, #width**
Rd is destination Reg.
**Rn** is source register.
**#lsb** starting bit from Rn.
**#width** number of bit starting from lsb of Rn.

**Example 10.11**  Copy 8 bits of R3 starting from bit 4 to R4; assume R3 contains 0x FFFFEBCD and R4 contains 0xEE035007.

BFI R4, R3, #4, #8, and this instruction will copy 8 bits from B4 to B11 of R3 into B0 through B7 of R4, the initial value of R3 in binary.

| 11101110000000 11010100000000111 |
|---|

The initial value of R4 in binary is

| 1110 1110 0000 0011 0101 0000 0000 0111 |
|---|

The instruction will copy 8 bits from bit 4 of R3 into R4 starting from bit 0 of R4.

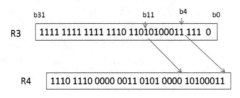

## 10.7   ARM Instruction Formats

Manufacture of processors will publish instruction format in order the assembler convert assembly language to machine code or binary. ARM instruction format is 32 bits for all types of instruction.

### 10.7.1   ARM Data Processing Instruction Format

The instruction format is used by assembler to convert instruction to machine code; Fig. 10.1 shows data processing instruction format.

#### 10.7.1.1   Condition Code

To determine if the instruction is a conditional or an unconditional instruction, Following are the code for condition

| 31 | 28 | 27 26 | 25 | 24 | | 21 | 20 | 19 | | 16 | 15 | | 12 | 11 | | 0 |
|---|---|---|---|---|---|---|---|---|---|---|---|---|---|---|---|---|
| Cond | | 0 0 | I | Op code | | | S | | Rn | | | RD | | | Operand 2 | |

**Fig. 10.1** Data processing instruction format

Condition code defines the type of instruction. If this field is set to 1110, then the instruction is an unconditional instruction, otherwise the instruction is a conditional instruction. To use an instruction as a conditional instruction, the condition will suffix to the instruction, the suffixes are as follows:

| Condition code | Condition | |
|---|---|---|
| 0000 | EQ | equal |
| 0001 | NE | not equal |
| 0010 | CS | carry set |
| 0111 | CC | carry is clear |
| 0100 | MI | negative (N flag is set) |
| 0101 | PL | positive (N flag is zero) |
| 0110 | VS | overflow set |
| 0111 | VC | overflow is clear |
| 1000 | HI | higher for unsigned number |
| 1001 | LS | less than for unsigned number |
| 1010 | GT | greater for signed number |
| 1011 | LT | signed less than |
| 1100 | GT | Greater Than |
| 1101 | LE | less than or equal |
| 1110 | AL | unconditional instructions |
| 1111 | Unused code | |

### 10.7.1.2   I bit

I = 0 means the operand2 is a register, I = 1 means the operand 2 is an immediate value.

### 10.7.1.3   Op Code

The OP Code determines types of instruction and following are the op codes for data processing instructions

| Instruction | Op code |
|---|---|
| AND | 0000 |
| EOR | 0001 |
| SUB | 0010 |
| RSB | 0011 |
| ADD | 0100 |
| ADC | 0101 |
| SBC | 0110 |

| 11 | | | 7 | 6 | 5 | | 4 | 3 | | 0 |
|---|---|---|---|---|---|---|---|---|---|---|
| # shift | | | | SH | | | 0 | Rm | | |

**Fig. 10.2** Operand2's format when bit 4 is equal to 0. **# Shift:** determines immediate value for number of times Rm must be shifted. **SH:** determines types of shift operation. Operation SH value: LSL 00 Logical Shift Left. LSR 01 Logical Shift Right. ASR 10 Arithmetic Shift Right. ROR 11 Rotate Right. **Rm:** second operand

| 11 | | 8 | 7 | 6 | 5 | 4 | 3 | | 0 |
|---|---|---|---|---|---|---|---|---|---|
| RS | | | 0 | SH | | 1 | Rm | | |

**Fig. 10.3** Format of Operand2 when bit 4 is equal to 1

| | | |
|---|---|---|
| RSC | 0111 | |
| TST | 1000 | |
| TEQ | 1001 | |
| CMP | 1010 | set condition by Op1-Op2 |
| CMN | 1011 | set condition for Op1+ Op2 |
| ORR | 1100 | |
| MOV | 1101 | Rd = operand2 |
| BIC | 1110 | |
| MVN | 1111 | Rd = NOT operand2 |

**S bit:** $S = 0$ do not change flag bits of PSR register, $S = 1$ set condition flags of PSR register.

**Rn:** Rn is first operand, and it can be any of the registers.

**Rd:** Rd is destination register, and it can be any of the registers.

**Operand2:** When $I = 0$ the operand2 is a register and Fig. 10.2 shows operand2's format.

**Example 10.12** Convert the following instructions to machine code.
    ADD R1, R2, R3, LSL #3.

| 31 | 28 | 27 26 | 25 | 24 | 21 | 20 | 19 | 16 | 15 | 12 | 11 | 7 | 6 | 5 | 4 | 3 | 0 |
|---|---|---|---|---|---|---|---|---|---|---|---|---|---|---|---|---|---|
| Cond | | 0 0 | I | Op code | | S | Rn | | RD | | #Shift | | SH | | | RM | |
| 1110 | | | 0 | 0100 | | 0 | 0010 | | 0001 | | 0011 | | 00 | | 0 | 0011 | |

When bit 4 of operand2 is set to 1, the number of times Rm must be shifted is in a register Rs as shown in Fig. 10.3.

I=1: The operand 2 would have the following format.

| 11 | 0 |
|---|---|
| Immediate Value | |

The following figure shows instruction formats for different type of data processing instructions.

| 31 30 29 28 | 27 26 25 | 24 23 22 21 20 | 19 18 17 16 | 15 14 13 12 | 11 10 9 8 7 | 6 5 | 4 | 3 2 1 0 |
|---|---|---|---|---|---|---|---|---|
| Condition | 0 0 0 | Opcode | S | Rd | Rn | # shift | SH | 0 | Rm |

Data Processing with Immediate Shift  ( ADD Rd, Rn, Rm, LSL, #4)

| 31 30 29 28 | 27 26 25 | 24 23 22 21 20 | 19 18 17 16 | 15 14 13 12 | 11 10 9 8 | 7 | 6 5 | 4 | 3 2 1 0 |
|---|---|---|---|---|---|---|---|---|---|
| Condition | 0 0 0 | Opcode | S | Rd | Rn | RS | 0 | SH | 1 | Rm |

Data Processing with Register Shift  ( ADD Rd, Rn, Rm, LSL, RS)

| 31 30 29 28 | 27 26 25 | 24 23 22 21 20 | 19 18 17 16 | 15 14 13 12 | 11 10 9 8 7 6 5 | 4 | 3 2 1 0 |
|---|---|---|---|---|---|---|---|
| Condition | 0 0 1 | Opcode | S | Rd | Rn | Immediate value |

Data Processing with immediate value  ( ADD Rd, Rn, #034 )

## 10.7.2  B and BL Instruction Format

Figure 10.4 shows instruction format for B{Cond} label and BL{Cond} subroutine.

## 10.7.3  Multiply Instruction Format

Multiply instruction offers MUL and MULA with the following forms:

(a) MUL{Cond}{s} Rd,Rm,Rs;Rd = Rm*Rs
(b) MLA{Cond}{s} Rd,Rm,Rs,Rn; Rd = Rm*Rs + Rn

Figure 10.5 shows instruction format for MUL and MLA.

| 31 | 28 27 | | 25 24 | 23 | 0 |
|---|---|---|---|---|---|
| Cond | 101 | | L | offset | |

**Fig. 10.4** Instruction format for B and BL instruction. L = 0 means Branch and condition for branch can be set by Cond field. L = 1 Mean Branch and Link

| 31 | 27 27 | 22 21 | 20 | 19 | 16 15 | 12 11 | 9 7 | 4 3 | 0 |
|---|---|---|---|---|---|---|---|---|---|
| Cond | 00000 | A | S | Rd | Rn | RS | 1001 | Rm | |

**Fig. 10.5** Instruction format for MUL and MLA. A = 0 MUL instruction. A = 1 MLA instruction. S = 0 Do not change flag bit. S = 1 Set the flag bits. Rd is destination register. Rs, Rm and Rn are the operands

## 10.7.4   Data Transfer Instructions (LDRB, LDR, STRB, and STR)

The data transfer instructions are used to transfer data from memory to registers and from registers to memory. Figure 10.6 shows data transfer for LDRB, LDR, STRB, and STR.

## 10.7.5   Data Transfer Half Word and Signed Number (LDRH, STRH, LDRSB, LDRSH)

Figure 10.7 shows instruction format for LDRH, STRH, LDRSB, and LDRSH.

| 31    28 | 27    25 | 24 | 23 | 22 | 21 | 20 | 19    16 | 15    12 | 11    0 |
|----------|----------|----|----|----|----|----|----------|----------|---------|
| Cond     | 0 1      | I  | P  | U  | B  | W  | L Rn     | Rd       | Offset  |

**Fig. 10.6** Instruction format for LDRB, LDR, STRB, and STR. **Rd:** Destination Register. **Rn:** Base Register. **L (load /Store):** L = 0 Store to memory, L = 1 Load from Memory. W = 0 no write back (keep Base Address the same value), W = 1 Modify Base address; write back (auto indexing). B = 0 transfer word, B = 1 transfer a byte. Up/Down bit; U = 0 subtract offset from base register. U = 1 add offset to the Base Register. P = 0 Post, add offset after transfer. P = 1 Pre, add offset before transfer.

I=0  offset is an immediate value                          11                         0

| Immediate Value |
|-----------------|

I =1   Offset is a register                          11          4    3          0

| Shift | Rm |
|-------|----|

| 31    27 | 28 25 | 24 23 | 22 | 21 | 20 | 19    16 | 15    12 | 11    8 | 7 | 6 | 5 | 4 | 3    0 |
|----------|-------|-------|----|----|----|----------|----------|---------|---|---|---|---|--------|
| Cond     | 000   | p     | U  | 0  | W  | L Rn     | RD       | 000     | 1 | S | H | 1 | RM     |

**Fig. 10.7** Instruction format for LDRH, STRH, LDRSB, and LDRSH. **Cond:** Condition. **P (Pre/Post indexing):** P = 0 means Post indexing and add or subtract from base register after transfer of data. P = 1 means Pre indexing and will add or subtract offset from base register before transfer of Data. U(**UP/Down**): U = 0 subtract offset from base register, U = 1 add offset to the base register. **W (Write Back)**: W = 0 no writeback, W = 1 write address to the base. **L (load/ Store):** L = 0 Store data into memory, L = 1 Read data from Memory. **Rn:** Base Register. **Rd:** destination Register. **SH:** determines is operation byte sign extension or half word or half word sign extension. SH = 00 means swap instruction. SH = 01 means unsigned halfword. SH = 10 Signed byte extension. SH = 11 Signed half word extension

### *10.7.6   Swap Memory and Register (SWAP)*

The xwap instruction combines the load and store instructions.

SWP Rd, Rm, [Rn]

 The register Rd is destination register, Rm is the source register, and Rn is base register.
 The Swap instruction perform the following functions

Rd memory [Rn] Load Rd from memory location [Rn]
[Rn] Rm store the contents of Rm in memory location [Rd]
   *SWPB Rd ,Rm, [Rn]*
  ;Swap one byte.

 Figure 10.8 shows swap instruction format

## 10.8   Summary

* ARM instruction uses LDR and STR to read and write to memory.
* The load instruction can be used to load one byte (LDRB), load 2 bytes (LDHB), and load 4 bytes (LDR).
* LDRSB (load signed extension) is used to load one byte and extended the sign of the data.
* LDRSH (load signed extension) is used to load two bytes and extended the sign of the data.
* The ARM pseudo instructions are ADR (load address of memory location) and LDR (load a 32 bit value into a register).
* ARM processor store instructions are STR (store one word), STRB (store one byte), and STRH (store half word).
* ARM offers several addressing modes and they are pre-indexed, pre-indexed with immediate offset, pre-indexed with register offset, pre-index with scaled register, pre-index with register offset and write back, post-index with immediate offset, post-index with register offset, and post-index with scaled register offset.
* Data can be represented in memory in the form of Big Endian and Little Endian.
* In Big Endian the most significant byte of a word is stored at the lowest address.
* In Little Endian the least significant byte of a word is stored at the lowest address.
* The instruction format for ARM 32-bit processor is 32 bits.

| 31   28 | 27   23 | 22 | 21 20 | 19   16 | 15   12 | 11   4 | 3   0 |
|---------|---------|----|-------|---------|---------|----------|-------|
| Cond    | 00010   | B  | 00    | Rn      | Rd      | 00001001 | Rm    |

**Fig. 10.8** Swap Instruction format

- 4 bits of ARM instruction format use to represent condition
- 4 bit of ARM instruction format is used to represent Operation Code (OP Code)
- The I bit in instruction format is used to define type of Operand2, if I = 0 the operand2 is register, if I = 1 mean operand 2 is an immediate data.
- The following figure shows Data Processing Instruction formats.
- Chapter 11 covers bitwise operations and conditional structures on C with corresponding ARM assembly language. Also covers ARM instruction format in order to convert assembly language to binary.

## Problems

1. Trace the following instructions; assume list starts at memory location 0x0000018 and using ARM Big Endian:
   ADRL R0, LIST   ; Load R0 with address of memory location list
      MOV R10, #0x2

   (a) LDR R1, [R0]
   (b) LDR R2, [R0, #4]!
   (c) LDRB R3, [R0], #1
   (d) LDRSB R5, [R0], #1
   (e) LDRSH R6, [R0]

      LIST DCB   0x34, 0xF5, 0x32, 0xE5, 0x01, 0x02, 0x8, 0xFE
2. Work problem #1 part a and b using Little Endian.

   (a) R1 = 0xE532F534
   (b) R2 = 0xFE080201

3. What is contents of register R7 after execution the following program?

ADRL R0, LIST
LDRSB R7, [R0]
LIST DCB 0xF5

4. What are the contents of register Ri for the following load Instructions? Assume R0 holds the address of list using Little Endian.

   (a) LDR R1, [R0]
   (b) LDRH R2, [R0]
   (c) LDRB R3, [R0], #1
   (d) LDRB R4, [R0]
   (e) LDRSB R5, [R0], #1
   (f) LDRSH R6, [R0]

      List DCB   0x34, 0xF5, 0x32, 0xE5, 0x01, 0x02.

5. The following memory is given, show the contents of each register, and assume R1 = 0x0001000 and R2 = 0x00000004 (use Little Endian).

   (a) LDR   R0, [R1]
   (b) LDR   R0, [R1, #4]
   (c) LDR   R0, [R1, R2]
   (d) LDR   R0, [R1, #4]!

   | | |
   |---|---|
   | 1000 | 23 |
   | | 13 |
   | | 56 |
   | | 00 |
   | 1004 | 45 |
   | | 11 |
   | | 21 |
   | | 88 |
   | 1008 | 03 |
   | | 08 |
   | | 35 |
   | | 89 |
   | 100C | 44 |
   | | 93 |

6. What are the effective address and contents of R5 after executing the following instructions? Assume R5 contains 0x 18 and r6 contains 0X00000020.

   (a) STR   R4, [R5]
   (b) STR   R4, [R5, #4]
   (c) STR   R4, [R5, #8]
   (d) STR   R4, [R5, R6]
   (e) STR   R4, [R5], #4

7. Write a program to add elements of List1 and store in the List2.

   List1 DCB 0x23, 0x45, 0x23, 0x11
   List2 DCB 0x0

8. Write a program to find the largest number and store it in memory location List3.

   List1 DCD 0x23456754
   List2 DCD 0x34555555
   List3 DCD 0x0

9. Write a program, find the sum of data in memory location LIST, and store the SUM in memory location sum using loop.

List DCB 0x23, 0x24, 0x67, 0x22, 0x99
SUM DCD 0x0

10. Write a program to read memory location LIST1 and LIST2 and store the sum in LIST3.

```
LIST1 DCD 0x00002345
    LIST2 DCD 0X00011111
    LIST3 DCD 0x0
```

11. Write a program to add eight numbers using Indirect addressing.

```
LIST DCB 0x5, 0x2,0x6,0x7 ,0x9,0x1,0x2,0x08
```

12. Write a program to add eight numbers using Post-index addressing.

```
LIST DCB 0x5, 0x2,0x6,0x7 ,0x9,0x1,0x2,0x08
```

13. What are the contents of R4 after execution of the following program.

```
__main
 LDR R1, =0xFF00FF
 ADRL R0, LIST1
 LDR R2, [R0]
 AND R4, R2, R1
 LIST1 DCD 0X45073487
```

14. Write a program to convert the following HLL to assembly language, assume R1 = 0x9, R2 = 0x6 and R3 = 0x5

```
If R1=R2 then
 R3= R3+1
 IF R1<R2 Then
 R3=R3-1
 If R1>R2 Then
 R3=R3-5
```

15. Write a subroutine to calculate value of Y where Y = X*2 + x + 5, assume x is represented by

```
LIST DCB 0x5
 LIST1 DCB 0x5
```

16. Write a program to compare two numbers and store largest number in a memory location LIST.

```
M1 EQU 5
 N1 EQU 6
 LIST2 DCB 0x0
```

17. What are content of registers for the following load instructions, assume R0 hold the address of list

    ADRL R0, list

    (a) LDRB R1, [R0, #0x1]
        R1=
    (b) LDRH R2, [R0, # 0x1] R2=
    (c) LDRB R3, [R0]
        R3=
    (d) LDRB R4, [R0 , #4]
        R4=
    (e) LDRH R5, [R0, #2]
        R5=
    (f) LDRH R6, [R0,#4]
        R6=

    List DCB 0x04, 0x05, 0x32, 0xE5, 0x01, 0x02

18. Convert the following ARM instruction to machine code

    (a) ADD R5, R6, R8
    (b) ADDNE R2, R3, 0x25
    (c) BNE label

19. Convert the following ARM instruction to machine code by hand then check you result by u µVision debugger

    (a) SUB R1, R2, R3, LSR #4
    (b) MOV R5, R6
    (c) LDR R4, [R0]

# Chapter 11
# Bitwise and Control Structures Used for Programming with C and ARM Assembly Language

**Objectives: After Completing this Chapter, you Should Be Able to**
- Write programs in both C and ARM assembly using AMD, OR, XOR, and NOT operations.
- Write programs capable of clearing and setting register bits in both C and ARM assembly.
- Write conditional statements such as If-Then, If-Then-Else, and loops in both C and ARM assembly.
- Write programs using Switch cases and While loops in both in C and ARM assembly.
- Learn the use and application of the C language in embedded programming.

## 11.1 Introduction

The C language is widely used for low level programming, meaning it gives the programmer more direct access to hardware and memory. Low level programming is needed for programming embedded systems, device drivers, assemblers, and operating systems. As it is necessary to access memory and specifically modify values at the bit level, C and ARM assembly provide a number of bitwise logic operations, while C allows for the utilization of control structures.

### 11.1.1 C Bitwise Operations

Table 11.1 shows the bitwise operations available in the C programming language, with corresponding operator symbols and examples. For the examples, assume A = 01001101 and B = 10101111

It possible to combine multiple bitwise operation in the same statement, such as:

A. Elahi, *Computer Systems*, https://doi.org/10.1007/978-3-030-93449-1_11

**Table 11.1** C language bitwise operations

| Operation | Symbol | Example |
|-----------|--------|---------|
| AND | & | C = A&B; <br> C = 00001101 |
| OR | \| | C = A\|B; <br> C = 11101111 |
| Exclusive OR | ^ | C = A^B; <br> C = 11100010 |
| NOT | ~ | C = ~A; <br> C = 10110010 |
| Shift Right | >> | C = A>> 2; // shift right A twice <br> C = 00010011 |
| Shift Left | << | C = B << 2; <br> C = 10111100 |

C = (B << 4) | (A & 0x0F);
C = 1010000 OR 00001100 = 10101100

### 11.1.1.1   Set a Bit of a Register to One

Changing a specific register bit to a one, also known as setting a bit, can be accomplished through a combination of OR and shifting operations. First, a single bit, set as a one, is shifted left K positions so that the one is moved into the same position of the bit to be set on register A. Then, the OR operation is performed using the value stored in register A and the new shifted value. Since the shifted value only contains a single one, the only modification of the register A value is the setting of that single bit. This can be generalized as:

A |= (1 << K); or A = A | (1 << K);

where k is the bit position in the register.

**Example 11.1**  Set bit a3 of an 8 bit register to one. Assume the bits are represented by $a_7$ $a_6$ $a_5$ $a_4$ $a_3$ $a_2$ $a_1$ $a_0$.

A |= 1 << 3;

The shifted value 1 << 3 evaluates to 1000, and the result of the OR operation becomes

$a_7$ $a_6$ $a_5$ $a_4$ $a_3$ $a_2$ $a_1$ $a_0$

+

0 0 0 0 1 0 0 0

-------------------------------------

$a_7$ $a_6$ $a_5$ $a_4$ 1 $a_2$ $a_1$ $a_0$

Bit $a_3$ becomes set as one, while the rest of the bits retain their original values.

**Example 11.2**  Set the bit position b3 of register x to one in C and ARM assembly

C programing          ARM assembly

```
int main(void){        AREA MYCODE, CODE, READONLY
    int x = 0;         EXPORT __main
    x | = (1 < < 3);   ENTRY
    return (0);        __main
}                      ; set b3 to one
                       MOV R1, #0X40
                       ORR R1, #0x08
                       STOP
                       B
                       STOP
                       END
```

### 11.1.1.2 Clear a Bit of a Register

Clearing a bit, or changing a bit to a zero, is similar to setting a bit. The bit position of the register bit to be cleared is again selected by shifting a one to the left N positions. However, this complement of this value is taken using the NOT operation, flipping the zeroes and ones. Then, bitwise AND is performed on the value from register A and the shifted and flipped value. The purpose of complementing.

$A = A \& (\sim (1 << n));$

or

$A \& = \sim (1 << n);$

**Example 11.3** Clear bit a5 of an 8 bit register X. Assume register X contains the value 0xFF.

$X \& = \sim (1 < < 5);$

The shifted value $1 < < 5$ evaluates to 100000, and the result of the NOT operation becomes

```
  ~
0 0 1 0 0 0 0 0
----------------------
1 1 0 1 1 1 1 1
```

The purpose of taking the complement is to ensure that the only bit is cleared from the register value is the bit in the desired position. When the AND operation is performed with the register value, the single zero in the complemented value will clear the desired bit, while the ones will ensure the rest of the register value remains unchanged, shown below:

```
1 1 1 1 1 1 1 1
  *
1 1 0 1 1 1 1 1
----------------------
1 1 0 1 1 1 1 1
```

Bit position 5 becomes cleared to zero, while the rest of the bits retain their original values.

**Example 11.4**  Clear bit position 5 of x = 0xFF in C and ARM assembly

```
C programming              ARM assembly
int main(void){            AREA    MYCODE, READONLY, CODE
   int x = 0xff;           EXPORT  __main
   x & = ~ (1 < < 5));     ENTRY
   return (0);             __main
}                          MOV R1, #0Xff
                           AND R1, #0xDF;  clear bit b5
                           stop    b  stop
                           END
```

## 11.2   Control Structures

Control structures within a programming language allow for a programmer to provide conditions within a program that can be used to repeat or jump to certain sections of code. Conditional statements can be used to execute code only when a condition is met, as well cause a block of statements to repeat. The amount of repetitions can be specified in a few ways, such as:

1. Until a condition is met
2. While a condition is true
3. Specified number of times

### 11.2.1   If-Then Structure

The If-Then structure is a conditional structure that will execute statements only condition is met (True or False), otherwise skipping those statements if the condition is not fulfilled. The C implementation of this structure can be seen below:

```
If (a < b) {
  statement1;
  statement2;
}
```

Statement 1 and Statement 2 are placed within the brackets of the If statement block, meaning that those statements will only be executed if the condition (a < b in

**Fig. 11.1**  If-Then Structure
Flowchart

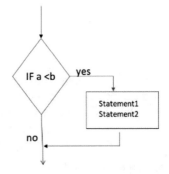

this case) is met. Otherwise, the program execution will skip those statements and continue normally. A flowchart of this structure can be seen in Fig. 11.1.

**Example 11.5**  If-Then Structure in C and ARM assembly language

```
C programming    ARM assembly
int main(void){  AREA MYCODE, CODE, READONLY
   int b = 15;   EXPORT __main
   int a = 5;    ENTRY
   int x = 2;    __main
   if (a < b) {  MOV R1, #0X5    ; R1 = a
      b++;       MOV R2, #0x15   ;R2 = b
      x = x + 1; MOV  R3, #2    ; R3 = x
   }             CMP R1, R2
   return(0);    ADDLT R3, R3, #0x01    ; X = X + 1
}                ADDLT R2, R2, #0x01    ; b = b + 1
                 stop    b  stop
                 END
```

## 11.2.2   If-Then-Else Structure

A modification of the previous structure is the If-Then-Else structure. Like the If-Then structure, If-Then-Else will execute a set of statements if condition true. However, if a condition is false a different set of statements can be specified before the program returns to normal execution.

```
If (a < b) {
   Statement1;
   Statement2;
}
```

```
Else {
  Statement3;
  Statement4;
}
```

Like the previous example, if the condition a < b is met, then Statement1 and Statement2 will be executed, and then the program will continue sequential execution. But, if the condition is false, then Statement3 and Statement4 will be executed instead, and again the program will return to normal execution. Since the condition can only be true or false, when one block is executed, the other block will be skipped.

**Example 11.6** If-Then-Else Structure in C and ARM assembly language

| C programming | ARM assembly |
|---|---|
| int main(void){ | AREA    MYCODE, READONLY, CODE |
| int i = 5; | EXPORT __main |
| int j = 6; | __main |
| int k; | MOV R0, #0x5    ;i = 5 |
| if (i < j) { | MOV R1, #0X6  ;j = 6 |
| k = i + j; | CMP R1,R2    ; compare |
| } | ADDLT R3, R2 , R1  ; If R1 < R2 add      ; R2 with R1 |
| else { | SUBGT R3, R2, R1  ; If R1 > R2   ;subtract |
| k = i - j; | done |
| } | B  done |
| return (0); | END |
| } | |

## 11.2.3  While Loop Structure

The While Loop will repeat a series of statements while a specified condition is still true, checking the status of the condition at the beginning of every iteration. If the condition is ever found to be false, then the enclosed statements will be skipped and the program will resume normal execution.

```
while (a < b)
{

  Statement1;
  Statement2;

}
```

If the condition is met, then Statement1 and Statement2 will execute. Instead of continuing, however, the while condition will be checked again. If it is true, the statements will repeat. But, if the statements cause the condition to become false at any point, the program will skip the enclosed code and continue execution.

**Example 11.7** While Loop Structure using C and ARM assembly language

```
C programming                        ARM assembly
//while Loop example                 AREA    MYCODE, READONLY, CODE
  int main (void) {                  EXPORT   __main
  /* local variable definition */    ENTRY
    int a = 10;                      __main
    int b = 20;                      MOV R0, #0x10
  /* while loop execution */         MOV R1, #0x20
    while(a < 15) {                  cc
      b = b - 2;                     CMP R0, #0x15
      a++;                           SUBLT R1,R1, #0x2
  }                                  ADDLT R0, R0, #1
  return (0);                        BLT    cc
}                                    done

                                     B   done
                                     END
```

## 11.2.4 For Loop Structure

The For Loop Structure also is used for repetition, but in this case the number of repetitions is specified. To do this, a new variable is initialized track the number of repetitions. A condition is then set using the new variable that will be used to test when the variable gets larger or smaller than a certain number. Finally, the variable must be set to update after each iteration, increasing or decreasing by a set amount to eventually fail the test condition and stop the repetitions. Figure 11.2 shows the flow chart for the For Loop structure.

**Example 11.8** For Loop structure using C and assembly

```
C programming                        ARM assembly
/* for Loop example */               AREA    MYCODE, READONLY, CODE
  int main (void) {                  EXPORT   __main
  /* local variable definition */    ENTRY
    int i;                           __main
    int a = 5;                       SUB R0, R0, R0; clear R0 = i
```

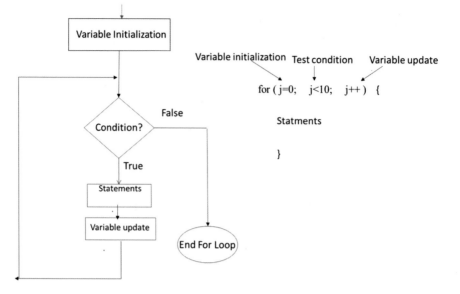

**Fig. 11.2** Flow chart of For Loop structure

```
for (i = 0; i < =10; i++){    MOV R1, #05    ; R1 = a
     a + = 1;                 cc
}                             CMP R0, # 10
   return 0;                  ADDLT  R1, #1
}                             ADD  R0, R0, #1
                              BLT cc
                              done

                              B    done
                              END
```

## 11.2.5  Switch Structure

The Switch structure is combination of multiple If-Then structures, separated into specific cases. A variable or expression is tested, and a matching case is chosen depending on the value. If no case matches the tested value, then a default case can be specified, acting like an Else structure. Only one case can match, but multiple cases can be grouped together to execute the same statements on a match. Figure 11.3 shows flowchart of switch structure.

**Fig. 11.3** Flowchart of switch structure

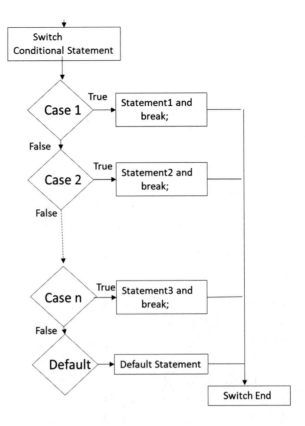

```
switch(x) {              // x is variable for testing

  case 0: statement1 //if x = 0

  case 1: statement2 //if x = 1

  case 2: statement3 //if x = 2
  .
  .
  default: statement  // for if x does not match a case
}
```

**Example 11.9** Writing a program to convert BCD to segment display using C and ARM assembly. Figure 11.4 shows the block diagram of a BCD to seven segment decoder and Table 11.2 shows the conversion table to display the proper decimal number.

In this program the input represented by decimal and output represented by hexadecimal. A Switch structure will check the input, and select the appropriate

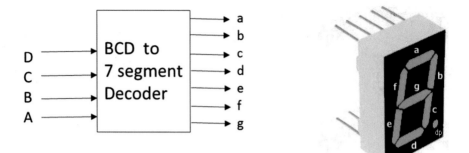

**Fig. 11.4** Block diagram of BCD to 7 segment decoder and &-segment display

**Table 11.2** Truth table for BCD to 7-segment decoder

| a | b | c | d | a | b | c | d | e | f | g | Display |
|---|---|---|---|---|---|---|---|---|---|---|---------|
| 0 | 0 | 0 | 0 | 1 | 1 | 1 | 1 | 1 | 1 | 0 | 0 |
| 0 | 0 | 0 | 1 | 0 | 1 | 1 | 0 | 0 | 0 | 0 | 1 |
| 0 | 0 | 1 | 0 | 1 | 1 | 0 | 1 | 1 | 0 | 1 | 2 |
| 0 | 0 | 1 | 1 | 1 | 1 | 1 | 1 | 0 | 0 | 1 | 3 |
| 0 | 1 | 0 | 0 | 0 | 1 | 1 | 0 | 0 | 1 | 1 | 4 |
| 0 | 1 | 0 | 1 | 1 | 0 | 1 | 1 | 0 | 1 | 1 | 5 |
| 0 | 1 | 1 | 0 | 1 | 0 | 1 | 1 | 1 | 1 | 1 | 6 |
| 0 | 1 | 1 | 1 | 1 | 1 | 1 | 0 | 0 | 0 | 0 | 7 |
| 1 | 0 | 0 | 0 | 1 | 1 | 1 | 1 | 1 | 1 | 1 | 8 |
| 1 | 0 | 0 | 1 | 1 | 1 | 1 | 0 | 0 | 1 | 1 | 9 |
| 1 | 0 | 1 | 0 | x | x | x | x | x | x | x | |
| 1 | 0 | 1 | 1 | x | x | x | x | x | x | x | |
| 1 | 1 | 0 | 0 | x | x | x | x | x | x | x | |
| 1 | 1 | 0 | 1 | x | x | x | x | x | x | x | |
| 1 | 1 | 1 | 0 | x | x | x | x | x | x | x | |
| 1 | 1 | 1 | 1 | x | x | x | x | x | x | x | |

case depending on the decimal number entered. Each case simply converts the decimal number to the hexadecimal value needed to light the segment display correctly.

| C programming | ARM assembly |
|---|---|
| int main () { | AREA     main, READONLY, CODE |
| /* local variable definition */ | EXPORT    __main |
| int input; | ENTRY |
| switch (input) { | __main |
| case 0: | MOV R2, #input |
| R1 = 0x37 / | MOV R1, R2 |
| break; | CMP R1, #0 |
| case 1 : | MOVEQ R3, #0x3F |
| R1 = 0x06 | BEQ  EXIT |
| break; | CMP  R1,# 1 |
| case 2 : | MOVEQ R3, #0x06 |
| R1 = 0x7D | BEQ  EXIT |
| break; | CMP R1, #2 |
| case 3 : | MOVEQ R3, #0x5B |
| R1 = 0x4f | BEQ  EXIT |
| break; | CMP R1,#0x03 |
| case 4 : | MOVEQ R3, #0x4F |
| R1 = 0x66 | BEQ  EXIT |
| break; | CMP  R1,# 4 |
| case 5 : | MOVEQ R3, #0x66 |
| R1 = 0x6d | BEQ  EXIT |
| case 6 : | CMP R1, #5 |
| R1 = 0x7D | MOVEQ R3, #0x6D |
| break; | BEQ  EXIT |
| case 7 : | CMP  R1,#0x06 |
| R1 = 0x07 | MOVEQ R3, #0x7D |
| break; | BEQ  EXIT |
| case 8 : | CMP R1, #7 |
| R1 = 0x7F | MOVEQ R3, #0x07 |
| break; | BEQ  EXIT |
| case 9 : | CMP  R1,#0x08 |
| R1 = 0x6F | MOVEQ R3, #0x7F |
| break; | BEQ  EXIT |
| default : | MOVEQ R3, #0x7D |
| R1 = 0x 7F | BEQ  EXIT |
| } | CMP R1, #9 |
| } | MOVEQ R3, #0x6F |
| | BEQ  done |
| | input     EQU     05 |
| | done |
| | |
| | B   done |
| | END |

**The *#define Preprocessor***

The define preprocessor directive can be used in a C program to define a constant value, with the following syntax:

```
#define identifier value;
```

**Example 11.10** Following program is used to calculate the area of rectangular using #define

```
C program                              Assembly
/* example of # define Preprocessor*/  AREA     main, READONLY, CODE
#define length  10                     EXPORT   __main
#define width  5                       ENTRY
int main (void) {                      __main
  /* local variable definition */      MOV R1, #length
  int area;                            MOV R2, #width
  area = length * width;               MUL R3, R1, R2
  return 0;                            length                EQU    05
}                                      width
                                       EQU    10
                                       stop

                                           B     stop
                                       END
```

## 11.3  ARM Memory Map

### 11.3.1  Introduction

ARM offers variety of the core processor based on their applications and they are:

**Cortex A series**: Cortex A series is a high performance processor for open operating system, the Cortex −A50 is a 64 bit process, application of Cortex-A series are Smart phones, Netbook, Digital TV, and eBook readers.

**Cortex −R series:** Cortex −R series is design for real-time application such as automobile braking, mass storage controller, printers, and networking.

**ARM Secure Processor**: This is an ultra-low power processor and it is used for SIMs cards, smart cards, and electronics passport. Figure 11.5 shows the general architecture of ARM processor.

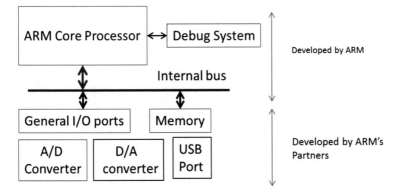

**Fig. 11.5**   Block Diagram of ARM processor with Peripherals

**ARM Cortex M series**: ARM Cortex M series is used as microcontroller for applications such as smart sensors, automobile control system, motor control, smart meters, and airbags. Figure 11.5 shows the general architecture of ARM processor. The ARM Corporation develop ARM core processor and ARM developer partners add more peripherals to the ARM processor such as A/D, D/A, CAN, Ethernet, and USB.

The Cortex –M3 is based on Harvard Architecture with 3 stage pipeline Architecture.

The ARM cortex is a low power processor and it is designed for embedded application with the following features, Fig. 11.6 shows ARM memory map and they are divided to multiple section and each assign specific function.

The Memory address from 0x00000000 to 0X1FFFFFFF use for storing code.

The Memory address from 0x20000000 to 0X3FFFFFFF use for storing Data.

The Memory address from 0x40000000 to 0X4FFFFFFF are used for memory mapped I/O several 8 bits.

The ARM Cortex Processor Contains Several Ports for input and output operations s and called General Purpose Input/Output (GPIO).

each port consist of 8 pins and user can configure the operation of each pin of the port such as input or output, this process involved multiple registers. ARM processors uses 4 memory locations as one register for memory mapped I/O, the following shows Registers with their corresponding address for port A.

| Register name | Offset | Function |
| --- | --- | --- |
| GPIOA_MODER | 0x40020000 | Pin direction/mode register |
| GPIOA_OTYPER | 0x40020004 | Pin output type register |
| GPIOA_OSPEEDR | 0x40020008 | Pin output speed register |
| GPIOA_PUPDR | 0x4002000C | Pull-up/pull-down register |
| GPIOA_IDR | 0x40020010 | Input data register |
| GPIOA_ODR | 0x40020014 | Output data register |

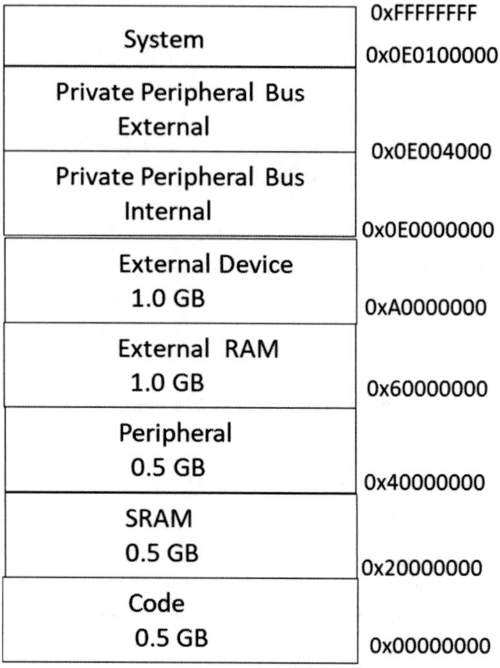

**Fig. 11.6** ARM memory
cortex -M3 Memory map

| | |
|---|---|
| System | OxFFFFFFFF |
| | 0x0E0100000 |
| Private Peripheral  Bus External | |
| | 0x0E004000 |
| Private Peripheral  Bus Internal | |
| | 0x0E0000000 |
| External Device 1.0 GB | |
| | 0xA0000000 |
| External  RAM 1.0 GB | |
| | 0x60000000 |
| Peripheral 0.5 GB | |
| | 0x40000000 |
| SRAM 0.5 GB | |
| | 0x20000000 |
| Code 0.5 GB | |
| | 0x00000000 |

As show in above table each memory mapped I/O takes for memory locations, following assembly language shows how to access each memory for reading and writing.

Example: write assembly language to read input register GPIOA_IDR and ADD 5 to it and send it to Output register GPIOA_ODR.

LDR R0, = GPOIA_IDR; load the address of GPOIA_IDR into R0
LDR R1, [R0]; read 4 bytes from memory 0x40020010 (address of input register)
ADD R1, R1, #05
LDR R0,= GPIOA_ODR; load address of output register
STR R1, [R0] ; Store R1 into GPIOA_ODR
GPOIA-IDR EQU 0x40020010
GPIOA_ODR EQU 0x40020014

## 11.4   Local and Global Variables

Local variables can be accessed only within the function or block in which they are defined and they are store in registers.

Consider C program in Fig. 11.7 where count is local variable, count assign to register R0 each time incremented by 2.

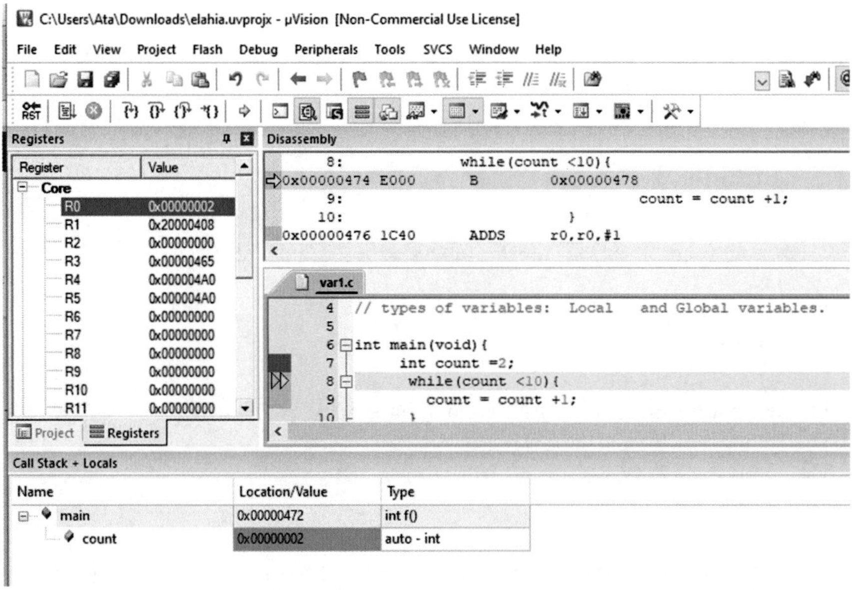

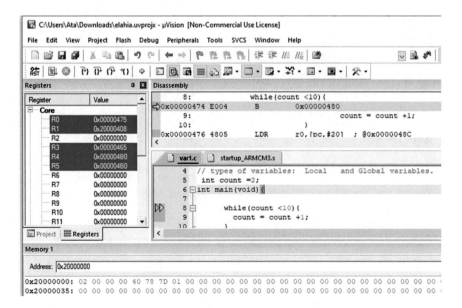

**Fig. 11.7**  C program with local variable

**Fig. 11.8**  C program with global variable

**Global Variables**: Global variables are variables declared outside a function and it is stored in memory and can be accessed by entire program, by moving count outside the main then count become global variable as shown in Fig. 11.8.

In Fig. 11.8 the count value is stored in memory location 0x200000 and each time goes through loop incremented by 1 and main program access this location and read count and store it into one of the registers.

## 11.5  Summary

- C programming used in most embedded programs, drivers, and operating systems.
- C programs offer several bitwise operations.
- The symbol for the bitwise AND operation in C is "&".
- The symbol for the bitwise OR operation in C is "|".
- The symbol for the bitwise AND operation in C is" ~".
- The symbol for logical shift right is ">> ".
- The symbol for logical shift left is "<< ".
- The symbol for XOR is ^.
- A C statement for setting bit position k in a register using is "A |= (1 << K)".
- A C statement for clearing bit position k in a register using is "A &= ~ (1 << n)".
- C has control structures that allow for the repetition or conditional execution of statements.
- The If-Then structure executes a set a statements only if a condition is met.
- The If-Then-Else structure executes a set of statement if a condition is met, or a different set of statements if it is not.
- The While Loop repeats a set of statements until a condition is false.
- The For Loop repeats a set of statements a set number of times.
- The Switch structure acts like a long series of If-Then statements.

## Problems

1. It recommended that student run all examples both in C and Assembly.
2. Write a program to set bit position 15 in C and assembly.
3. Write a program to set bit passion b5b4 to 11.
4. Write a program to clear position 15 in C and assembly.
5. Write a program to shift right contents of register 8 times in C and Assembly, assume register holds 0x400.
6. Write a program to shift left contents register 8 times in C and Assembly, assume register holds 0x4.
7. Write a program in C and assembly to find the largest number of following data.

   0x2, 0x5, 0x45, 0x24.

8. Write a switch statement to convert decimal number to ASCII using C and assembly.
9. Write a program two swap high digit with low digit of 0x45 using C and Assembly.

# Appendix A: List of Digital Design Laboratory Experiments Using LOGISIM

**Appendix A: List of the Digital Design Lab Experiments Using Logisim**

Lab#1: Introduction to Logisim
Lab#2: Logic Gates I
Lab #3: Logic Gates II
Lab #4: Combinational Logic Circuit
Lab #5: Decoder
Lab#6: 4-bit Binary Adder
Lab #7: 4*1 Multiplexer
Lab#8: BCD to 7 Segment Decoder
Lab #9: Design a 4-bit Arithmetic Logic Unit (ALU)
Lab #10: S-R latch, D –Flip Flop and Register Operations
Lab #11: J-K Flip Flop and T-Flip Flop Operations
Lab #12: Shift Register Operation
Lab # 13: Register Transfer Operation
Lab #14: Designing Counter
Lab #15: Random Access Memory (RAM)

# Appendix B: Solution to the Even Problems

## Chapter 1: Problems and Questions

2. List types of Computer?

   PC, Server, Embedded system , supercomputer, Cloud computer and PMD

4. List the two Operating Systems

   Windows and Linux

6. List three computer output devices.

   Monitor, speaker, printer

8. Show a digital signal.

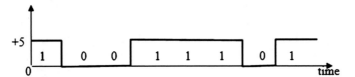

10. How many bits is?

    (a) Byte = 8 bits
    (b) Half word = 16 bits
    (c) Word = 32 bits

12. Convert the following binary to decimal

    (a) 111111
    $$2^6 - 1 = 63$$
    (b) 10 10101
    $$64 + 0 + 16 + 4 + 1 = 85$$
    (c) 1101001.101

$$64 + 32 + 0 + 16 + 0 + 0 + 1 + 1/2 + 1/8 = 97.75$$

14. Convert the following binary numbers to hexadecimal

    (a) $1110011010 = 39A$
    (b) $1000100111 = 227$
    (c) $101111.101 = 101111.1010 = 2F.A$

16. The following frequencies of digital signal are given, find clock cycle of digital signal

    (a) 10 Hz T $= 1/F = 1/10 = 0.1$ s
    (b) 200 Hz T $= 1/F = 1/200 = 0.005$
    (c) 10000 Hz T $= 1/10000 = 0.1$ ms
    (d) 4 MHz T $= 1/4*10^6 = 0.25*10^{-6} = 0.25$ μs

18. Convert each of the following number to base 10

    (a) $(34A)_{16} = A*16^0 + 4*16^1 + 3*16^2 = 1 + 64 + 3*256 = 842$
    (b) $(FAC)_{16} = C*16^0 + A*16^1 + F*16^2 = 12*1 + 10*16 + 15*256 = 4012$

20. Perform the following additions

    ```
    1101010
      1100101
    +1011011
        +1010111
    ----------

    ----------
    11000101
      10111100
    ```

22. The word "LOGIC" is given

    (a) Represent in ASCCII
    (b) Add even parity bit to each character and represent each character in hex
        L $= 1001100$
        O $= 1001111$
        G $= 1001010$
        I $= 1001001$
        C $= 1000011$
    (c) L $= 11001100$ O $= 11001111$ G $= 11001010$ I $= I1001001$ C $= 11000011$
        L $=$ CC O $=$ CF G $=$ CA I $=$ C9 C $=$ C3

24. Represent $(100101100111)_{BCD}$ in decimal
    $(967)_{10}$

26. Convert the following two's complement numbers to decimal:

(a) 1011
(b) 11111001
(c) 10011111

The most significant bit represents sign which is negative, the two's complement of number without sign

(a) 1011 = 011, 100 + 1 = 101, the number 1011 = −5
(b) 11111001

− 1111001 → 0000110 + 1 = 0000111 = −7

(c) 10011111

− 0011111 two's complement 1100000 + 1 = −1100001 = −97

28. Perform addition of the following signed numbers assume each number represented by 6 bits and state if result of each addition produce overflow

(a) (+12) + (+7)
12 in binary is 1100
+12 in 6 bits is 0 01100 , the most significant is sign bit
+7 in 6 bits 0 00 111
(+12) = 001100
(+7) = 000111
------------------------
010011 = +19
(b) (+25) + (+34)
(+25) = 011001
(+30) = 011110
----------
110111 adding two positive number results negative number, therefor it is call carry overflow
(c) (−5) + (+9)
−5 in signed two's complement
−5 in signed magnitude 100101

Ignore the sign and find two's complement 00101
Two's complement of 00110 = 11010 + 1 = 11011
Add sign to 11010 results 111011

−5 = 111011
+9 = 001001
------------
1 00100 , discard the carry and result is 000100 which is +4
(d) (−6) + (−7)
−6 in signed two's complement is 111010
−7 in signed two's complement is 111001

$(-6) = 111010$

$+(-7) = 111001$

-------------

1 110011 adding two negative numbers and result is negative, discard the carry and result 110011 in signed two's complement, ignore the sign and find two's complement of 10011

Two's complement of $10011 = 01100 + 1 = 01101$

Signe is negative and result is $-13$

30. Represent the following decimal number in IEEE745 single precision

(a) 34.375

(b) $-0.045$

(a) $34.37 = (100010.0101111)_2$

$1.000100101111*2^{+5}$

Exponent $= +5 + 127 = 132 = 10000100$

S Exponent

Mantissa

0 10000100 00010010111100000000000

(b) $-0.045 = -(0.0111) = -1.11*2^{-2}$

Exponent $= -2 + 127 = 125$

S Exponent

Mantissa

1 01111101 11000000000000000000000

32. List the types of transmission modes

Serial Transmission

Parallel Transmission

34. Represent each of the following numbers in 8-bit signed two's complement

(a) $-15$

(b) $-24$

(c) $-8$

(a) $-15$ in signed magnitude is 10001111

$-15$ in signed two's complement is magnitude is 11110001

(b) $-24$ in signed magnitude is 10011000

$-24$ in signed two's complement is 11101000

(c) $-8$ in signed magnitude is 10001000

(d) $-8$ in signed two's complement is 111010000

# Chapter 2: Answers

2. If A = 11001011 and B = 10101110, then what are the results of the following operations value? of the following operations?

   (a) A AND B
   (b) A OR B

   (a) Performing bit by bit and operation

   > A = 11001011
   > B = 10101110
   > A AND B = 10001010

   (b) Performing bit by bit or operation

   > A = 11001011
   > B = 10101110
   > A OR B = 11101111

4. Draw logic circuits for the following functions:

   (a) $F(X,Y,Z) = XY' + YZ + XZ'$
   (b) $F(X,Y,Z) = (X + Y')(Y + Z)(X' + Z')$

   (a)

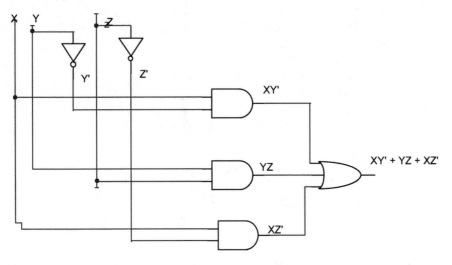

   (b)

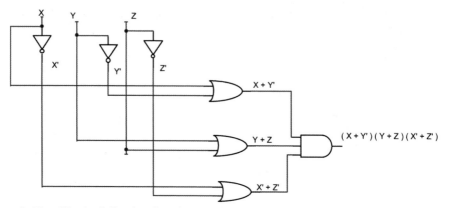

6. Simplify the following functions

(a) $F(X, Y, Z) = XY + X'Y + XZ$
   Solution
   $F(X, Y, Z) = Y(X + X') + XZ = Y + XZ$

(b) $F(X, Y, Z) = (X + Y)(X' + Y + Z)$
   Solution
   $F(X, Y, Z) = (XX' + XY + XZ + X'Y + YY + YZ)$ where $XX' = 0$ and
   $YY = Y$
   $F(X, Y, Z) = (XY + XZ + X'Y + Y + YZ)$
   $F(X, Y, Z) = Y(X + X' + 1 + Z) + XZ$
   $F(X, Y, Z) = Y + XZ$
   $F(X, Y, Z) = XY'Z + XYZ + Y'Z F(X, Y, Z) = XZ(Y' + Y) + Y'Z$
   $F(X, Y, Z) = XZ + Y'Z$

(c) $F(X, Y, Z) = XY + YX'Z$
   $F(X, Y, Z) = Y(X + X'Z)$ where $X + X'Z = X + Z$
   $F(X, Y, Z) = Y(X + Z)$

(d) $F(X, Y, Z) = X'Y + YXZ'$
   $F(X, Y, Z) = Y(X' + XZ') = Y(X' + Z')$

(e) $F(X, Y, Z) = XY + (X + Y + Z)'X + YZ$
   $F(X, Y, Z) = XY + (X'Y'Z')X + YZ$
   $F(X, Y, Z) = XY + YZ$

(f) $F(X, Y, Z) = (XY)' + (X' + Y + Z')' YZ$
   $F(X, Y, Z) = XY + (X'Y'Z')X + YZ$
   $F(X, Y, Z) = XY + YZ$

(g) **$F(X, Y, Z) = (XY)' + (X' + Y + Z')'$**
   $F(X, Y, Z) = X' + Y' + XY'Z$
   $F(X, Y, Z) = Y'(1 + XZ) + X' = X' + Y'$

8. If A = 10110110 and B = 10110011, then find

   (a) A NAND B
   (b) A NOR B
   (c) A XOR B

   (a) NAND each bit of A with corresponding bit of B

      A = 10110110
      B = 10110011
      A NAND B = 01001101

   (b) A = 10110110
      B = 10110011
      A NOR B = 01001000

   (c) A = 10110110
      B = 10110011
      A XOR B = 00000101

10. Show the output of the following logic circuits:

(a)

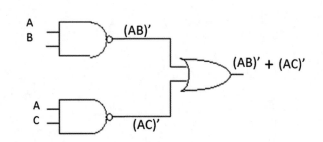

(b)

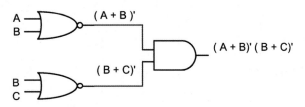

(c)

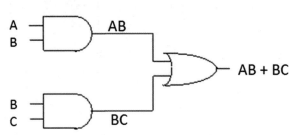

12. Find the output function of the following logic circuit:

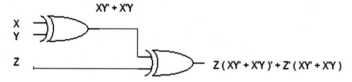

14. Show the truth table for each of the following functions:

(a) $F(X,Y,Z) = XY' + XZ' + YZ$
(b) $F(X,Y,Z) = (X + Y)(X + Z')$
(c) $F(X,Y,Z) = XY(Y + Z')$

(a)

| X | Y | Z | Y' | (XY) | (XY)' | (X + Y') | X (X + Y') | (XY)' + X (X + Y') |
|---|---|---|----|------|-------|----------|------------|--------------------|
| 0 | 0 | 0 | 1  | 0    | 1     | 1        | 0          | 1                  |
| 0 | 0 | 1 | 1  | 0    | 1     | 1        | 0          | 1                  |
| 0 | 1 | 0 | 0  | 0    | 1     | 0        | 0          | 0                  |
| 0 | 1 | 1 | 0  | 0    | 1     | 0        | 0          | 0                  |
| 1 | 0 | 0 | 1  | 0    | 1     | 1        | 1          | 1                  |
| 1 | 0 | 1 | 1  | 0    | 1     | 1        | 1          | 1                  |
| 1 | 1 | 0 | 0  | 1    | 0     | 1        | 1          | 1                  |
| 1 | 1 | 1 | 0  | 1    | 0     | 1        | 1          | 1                  |

(b) $F(X,Y,Z) = (X + Y + Z')'(X' + Y')$

| X | Y | Z | X' | Y' | Z' | (X' + Y') | (X + Y + Z') | (X + Y + Z')' | (X + Y + Z')'(X' + Y') |
|---|---|---|----|----|----|-----------|--------------|---------------|------------------------|
| 0 | 0 | 0 | 1  | 1  | 1  | 1         | 1            | 0             | 0                      |
| 0 | 0 | 1 | 1  | 1  | 0  | 1         | 0            | 1             | 1                      |
| 0 | 1 | 0 | 1  | 0  | 1  | 1         | 1            | 0             | 0                      |
| 0 | 1 | 1 | 1  | 0  | 0  | 1         | 1            | 0             | 0                      |
| 1 | 0 | 0 | 0  | 1  | 1  | 1         | 1            | 0             | 0                      |
| 1 | 0 | 1 | 0  | 1  | 0  | 1         | 1            | 0             | 0                      |
| 1 | 1 | 0 | 0  | 0  | 1  | 0         | 1            | 0             | 0                      |
| 1 | 1 | 1 | 0  | 0  | 0  | 0         | 1            | 0             | 0                      |

(c)

| X | Y | Z | Y' | (X XOR Y) | (X NOR Y') | (X XOR Y) (X NOR Y') |
|---|---|---|----|-----------|------------|----------------------|
| 0 | 0 | 0 | 1  | 0         | 0          | 0                    |
| 0 | 0 | 1 | 1  | 0         | 0          | 0                    |
| 0 | 1 | 0 | 0  | 1         | 1          | 1                    |
| 0 | 1 | 1 | 0  | 1         | 1          | 1                    |
| 1 | 0 | 0 | 1  | 1         | 0          | 0                    |
| 1 | 0 | 1 | 1  | 1         | 0          | 0                    |
| 1 | 1 | 0 | 0  | 0         | 0          | 0                    |
| 1 | 1 | 1 | 0  | 0         | 0          | 0                    |

(d)

| X | Y | Z | X′ | Y′ | (X′ + Y′ + Z) | (X + Y) | (X′ + Y′ + Z) (X + Y) |
|---|---|---|----|----|---------------|---------|------------------------|
| 0 | 0 | 0 | 1 | 1 | 1 | 0 | 0 |
| 0 | 0 | 1 | 1 | 1 | 1 | 0 | 0 |
| 0 | 1 | 0 | 1 | 0 | 1 | 1 | 1 |
| 0 | 1 | 1 | 1 | 0 | 1 | 1 | 1 |
| 1 | 0 | 0 | 0 | 1 | 1 | 1 | 1 |
| 1 | 0 | 1 | 0 | 1 | 1 | 1 | 1 |
| 1 | 1 | 0 | 0 | 0 | 0 | 1 | 0 |
| 1 | 1 | 1 | 0 | 0 | 1 | 1 | 1 |

16. Draw logic circuits for the following functions.

   (a) $F(X,Y,Z) = (X + Y)' + YZ$
   (b) $F(X,Y,Z) = (XYZ)' + XZ + YZ$

   (a)

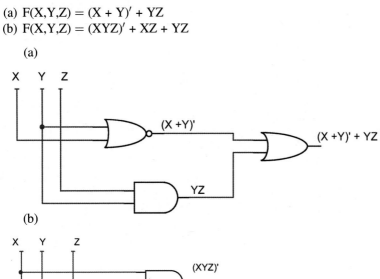

   (b)

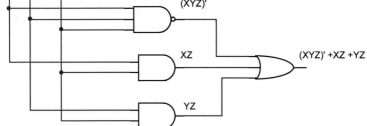

# Chapter 3: Solution

Problems

2. Generate truth tables for the following functions:

(a) $F(X,Y,Z) = \Sigma(1,3,6,7)$

| X | Y | Z | F |
|---|---|---|---|
| 0 | 0 | 0 | 0 |
| 0 | 0 | 1 | 1 |
| 0 | 1 | 0 | 0 |
| 0 | 1 | 1 | 1 |
| 1 | 0 | 0 | 0 |
| 1 | 0 | 1 | 1 |
| 1 | 1 | 0 | 0 |
| 1 | 1 | 1 | 1 |

(b) $F(X,Y,Z) = \pi(1,3,4)$

Maxterms represent zeros in the truth table.

| X | Y | Z | F |
|---|---|---|---|
| 0 | 0 | 0 | 1 |
| 0 | 0 | 1 | 0 |
| 0 | 1 | 0 | 1 |
| 0 | 1 | 1 | 0 |
| 1 | 0 | 0 | 0 |
| 1 | 0 | 1 | 1 |
| 1 | 1 | 0 | 1 |
| 1 | 1 | 1 | 1 |

(c) $F(W,X,Y,Z) = \Sigma(1,4,7,10,12,15)$

Minterms represent ones in the truth table.

| W X Y Z | F |
|---------|---|
| 0 0 0 0 | 1 |
| 0 0 0 1 | 0 |
| 0 0 1 0 | 0 |
| 0 0 1 1 | 0 |
| 0 1 0 0 | 1 |
| 0 1 0 1 | 0 |
| 0 1 1 0 | 0 |
| 0 1 1 1 | 0 |
| 1 0 0 0 | 0 |
| 1 0 0 1 | 0 |
| 1 0 1 0 | 1 |
| 1 0 1 1 | 0 |
| 1 1 0 0 | 0 |
| 1 1 0 1 | 0 |
| 1 1 1 0 | 0 |
| 1 1 1 1 | 1 |

(d)  $F(W,X,Y,Z) = \pi(2,3,4,7,10,11,12,13)$
   Maxterms represent zeros in the truth table.
   $F(X,Y,Z) = Y'Z' + YZ$

4. Simplify the following functions using K-map

(a)

| X | Y | F |
|---|---|---|
| 0 | 0 | 1 |
| 0 | 1 | 1 |
| 1 | 0 | 1 |
| 1 | 1 | 0 |

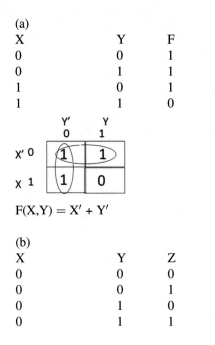

$F(X,Y) = X' + Y'$

(b)

| X | Y | Z | F |
|---|---|---|---|
| 0 | 0 | 0 | 1 |
| 0 | 0 | 1 | 1 |
| 0 | 1 | 0 | 0 |
| 0 | 1 | 1 | 1 |

| 1 | 0 | 0 | 0 |
| 1 | 0 | 1 | 1 |
| 1 | 1 | 0 | 1 |
| 1 | 1 | 1 | 0 |

|  | Y′ |  | Y |  |
|---|---|---|---|---|
|  | 00 | 01 | 11 | 10 |
| X′ 0 | 1 | 1 | 1 | 0 |
| X 1 | 0 | 1 | 0 | 1 |
|  | Z′ | Z | Z′ |  |

$$F(X,Y,Z) = X'Y' + X'Z + Y'Z + XYZ'$$

(c)

| A | B | C | D | F |
|---|---|---|---|---|
| 0 | 0 | 0 | 0 | 1 |
| 0 | 0 | 0 | 1 | 0 |
| 0 | 0 | 1 | 0 | 1 |
| 0 | 0 | 1 | 1 | 1 |
| 0 | 1 | 0 | 0 | 0 |
| 0 | 1 | 0 | 1 | 1 |
| 0 | 1 | 1 | 0 | 1 |
| 0 | 1 | 1 | 1 | 1 |
| 1 | 0 | 0 | 0 | 1 |
| 1 | 0 | 0 | 1 | 1 |
| 1 | 0 | 1 | 0 | 0 |
| 1 | 0 | 1 | 1 | 0 |
| 1 | 1 | 0 | 0 | 0 |
| 1 | 1 | 0 | 1 | 1 |
| 1 | 1 | 1 | 0 | 1 |
| 1 | 1 | 1 | 1 | 1 |

$$F(W,X,Y,Z) = WZ + W'X'Y' + W'XZ' \quad \text{SOP}$$

6. Simplify the following functions where D is a *don't care* function:

(a) $F(X,Y,Z) = \sum(0, 3, 4)$
$D(X,Y,Z) = \sum(2, 6)$

|  | Y′ |  | Y |  |
|---|---|---|---|---|
|  | 00 | 01 | 11 | 10 |
| X′ 0 | 1 | 0 | 1 | d |
| X 1 | 1 | 0 | 0 | d |
|  | Z′ | Z | Z′ |  |

$$F(X,Y,Z) = Z' + X'Y$$

(b) $F(W,X,Y,Z) = \sum(0, 1, 3, 5, 9, 11)$
$D(W,X,Y,Z) = \sum(2, 4, 8, 10)$

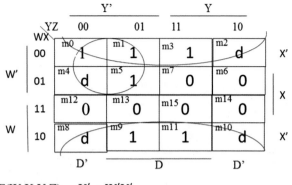

$$F(W,X,Y,Z) = X' + W'Y'$$

8. Simplify the following function and draw logic circuit using

    (a) NAND gates
    (b) NOR gates

$$F(W,X,Y,Z) = W'X'Z' + XY'Z' + WX + WY + WY'X'Z'$$

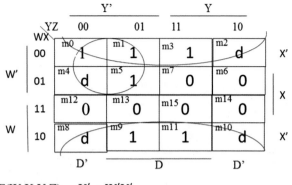

$$F(W,X,Y,Z) = X'Z' + WX + Y'Z' + WY$$
$$F(W,X,Y,Z) = [(X'Z' + WX + Y'Z' + WY)']'$$
$$F(W,X,Y,Z) = [(X'Z')'\ (WX)'\ (Y'Z')'\ (WY)']'\quad \text{NAND form}$$

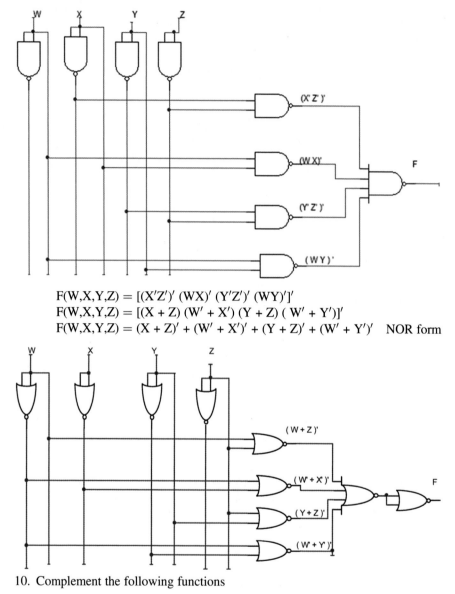

$$F(W,X,Y,Z) = [(X'Z')' (WX)' (Y'Z')' (WY)']'$$
$$F(W,X,Y,Z) = [(X + Z) (W' + X') (Y + Z) ( W' + Y')]'$$
$$F(W,X,Y,Z) = (X + Z)' + (W' + X')' + (Y + Z)' + (W' + Y')' \quad \text{NOR form}$$

10. Complement the following functions

(a) $F(X, Y, Z) = (X' + Y)(X + Z)(Y' + Z')$

$$F'(X, Y, Z) = [(X' + Y)(X + Z)(Y' + Z')]'$$
$$F'(X, Y, Z) = [(X' + Y)' + (X + Z)' + (Y' + Z')'$$
$$F'(X, Y, Z) = (x')'(y)' + (X)'(Z)' + (y')'(z')'$$
$$F'(X, Y, Z) = XY + XZ + YZ$$

(b) $F(X, Y, Z) = X'Y + XY'Z + XYZ'$

$$F'(X, Y, Z) = (X'Y + XY'Z + XYZ')'$$
$$F'(X, Y, Z) = (X'Y)' \, (XY'Z)' \, (XYZ')'$$
$$F'(X, Y, Z) = ((X')' + (Y)') \, ((X)' + (Y')' + (Z)') \, ((X)' + (Y)' + (Z')')$$
$$F'(X,Y,Z) = (X + Y')(X' + Y + Z') \, (X' + Y' + Z)$$

## Chapter 4

2. Find the output of the following gates:

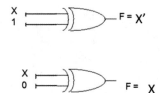

4. Design a logic circuit with three inputs and one output; the output generates even parity bit of the inputs; assume zero is even.

   (a) Show the truth table.
   (b) Find output function.
   (c) Draw logic circuit.

| X | Y | Z | F |
|---|---|---|---|
| 0 | 0 | 0 | 1 |
| 0 | 0 | 1 | 0 |
| 0 | 1 | 0 | 0 |
| 0 | 1 | 1 | 1 |
| 1 | 0 | 0 | 0 |
| 1 | 0 | 1 | 1 |
| 1 | 1 | 0 | 1 |
| 1 | 1 | 1 | 0 |

$$F(X,Y,Z) = m0 + m3 + m5 + m6$$

| | Y' | | Y | |
|---|---|---|---|---|
| | 00 | 01 | 11 | 10 |
| X' 0 | 1 | 0 | 1 | 0 |
| X 1 | 0 | 1 | 0 | 1 |

$$F(X,Y,Z) = X'Y'Z' + X'YZ + XY'Z + XYZ' = X'(Y'Z' + YZ) + X \, (Y'Z + YZ')$$
If   $Y'Z + YZ' = A$, then $Y'Z' + YZ = A'$
Therefore function F can be written as
$$F(X,Y,Z) = X'A' + XA = X \text{ XNOR } A$$

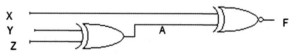

6. Implement the following functions using only one decoder and external gates:

$$F1(X,Y,Z) = \Sigma(0, 3, 4)$$
$$F2(X,Y,Z) = \Sigma(2, 3, 5)$$

The function requires a 3*8 decoder.

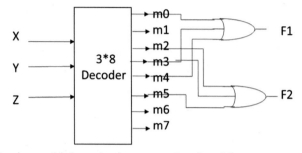

8. The following multiplexer is given; complete its table.

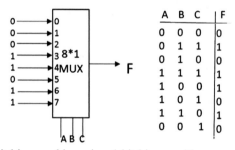

| A | B | C | F |
|---|---|---|---|
| 0 | 0 | 0 | 0 |
| 0 | 1 | 1 | 1 |
| 0 | 1 | 0 | 0 |
| 1 | 1 | 1 | 1 |
| 1 | 0 | 0 | 1 |
| 1 | 0 | 1 | 0 |
| 1 | 1 | 0 | 1 |
| 0 | 0 | 1 | 0 |

10. Design an 8-bit binary adder using 4-bit binary adders.

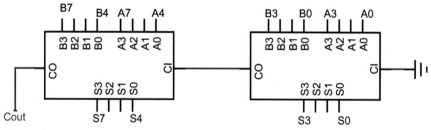

12. Design a combination logic with three inputs and three outputs, if input 0, 1, 2, or 3, then output 3 more than input, if input 4, 5, 6, or 7 then output 3 less than input.

| X | Y | Z | A | B | C |
|---|---|---|---|---|---|
| 0 | 0 | 0 | 0 | 1 | 1 |
| 0 | 0 | 1 | 1 | 0 | 0 |
| 0 | 1 | 0 | 1 | 0 | 1 |

| 0 | 1 | 1 | 1 | 1 | 0 |
|---|---|---|---|---|---|
| 1 | 0 | 0 | 0 | 0 | 1 |
| 1 | 0 | 1 | 0 | 1 | 0 |
| 1 | 1 | 0 | 0 | 1 | 1 |
| 1 | 1 | 1 | 1 | 0 | 0 |

Making K-map for A, B, and C

**K-map for A**

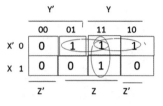

$A(X,Y,Z) = X'Z + X'Y + YZ$

**K-map for B**

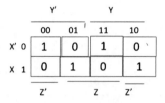

$B(X,Y,Z) = X'Y'Z' + X'YZ + XY'Z + XYZ' = X'(Y'Z' + YZ) + X(Y'Z + YZ')$

If $Y'Z + YZ' = W$ then

$B(X,Y,Z) = X'W' + XW = X$ XNOR $W$ and $W = Y$ XOR $Z$

Function for C—by looking at the truth table, the column for C is complement of Z, then

$C = Z'$

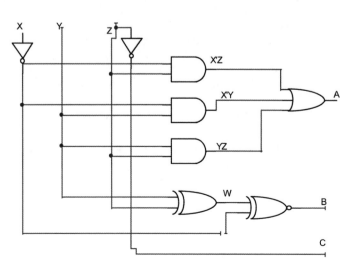

14. Design a combinational circuit with four inputs and one output; the input to the combination circuit is BCD, and the output generates even parity for the input.

| W | X | Y | Z | F |
|---|---|---|---|---|
| 0 | 0 | 0 | 0 | 0 |
| 0 | 0 | 0 | 1 | 1 |
| 0 | 0 | 1 | 0 | 1 |
| 0 | 0 | 1 | 1 | 0 |
| 0 | 1 | 0 | 0 | 1 |
| 0 | 1 | 0 | 1 | 0 |
| 0 | 1 | 1 | 0 | 0 |
| 0 | 1 | 1 | 1 | 1 |
| 1 | 0 | 0 | 0 | 1 |
| 1 | 0 | 0 | 1 | 0 |
| 1 | 0 | 1 | 0 | d |
| 1 | 0 | 1 | 1 | d |
| 1 | 1 | 0 | 0 | d |
| 1 | 1 | 0 | 1 | d |
| 1 | 1 | 1 | 0 | d |
| 1 | 1 | 1 | 1 | d |

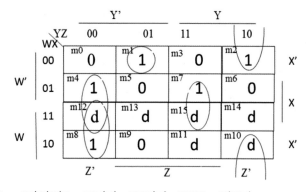

$$F(W,X,Y,Z) = W'X'Y'Z + XY'Z' + WY'Z' + XYZ + X'YZ'$$

16. Design a 4-bit ALU to perform the following functions:
A+B, A−B, A+1, A', B', A OR B, A XOR B, A AND B
Solution
This is 4-bit ALU; therefore, it requires four multiplexers. This ALU has eight functions; therefore, each multiplexer is 8*1.

18. Find the output F for the following combinational logic:

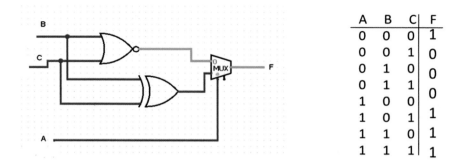

| A | B | C | F |
|---|---|---|---|
| 0 | 0 | 0 | 1 |
| 0 | 0 | 1 | 0 |
| 0 | 1 | 0 | 0 |
| 0 | 1 | 1 | 0 |
| 1 | 0 | 0 | 0 |
| 1 | 0 | 1 | 1 |
| 1 | 1 | 0 | 1 |
| 1 | 1 | 1 | 1 |

## Chapter 5: Problems

2. Complete the following table for JK flip-flop:

| J | K | Q(t) present output | Q(t + 1) next output |
|---|---|---|---|
| 0 | 0 | 0 | 0 |
| 0 | 0 | 1 | 1 |
| 0 | 1 | 0 | 0 |
| 0 | 1 | 1 | 0 |
| 1 | 0 | 0 | 1 |
| 1 | 0 | 1 | 1 |
| 1 | 1 | 0 | 1 |
| 1 | 1 | 1 | 0 |

4. The following figure shows a sequential logic; complete the following table; assume initial value of Q1 = 0 and Q2 = 0. Use logicism to verify your answer.

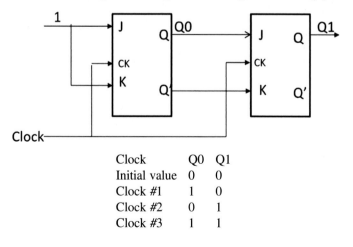

| Clock | Q0 | Q1 |
|---|---|---|
| Initial value | 0 | 0 |
| Clock #1 | 1 | 0 |
| Clock #2 | 0 | 1 |
| Clock #3 | 1 | 1 |

6. The following shift register is given; find the output after five clock pulses.

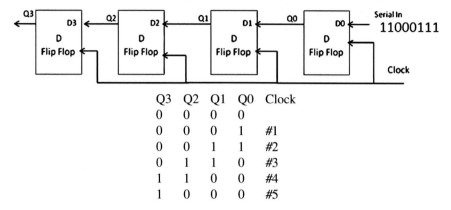

| Q3 | Q2 | Q1 | Q0 | Clock |
|---|---|---|---|---|
| 0 | 0 | 0 | 0 | |
| 0 | 0 | 0 | 1 | #1 |
| 0 | 0 | 1 | 1 | #2 |
| 0 | 1 | 1 | 0 | #3 |
| 1 | 1 | 0 | 0 | #4 |
| 1 | 0 | 0 | 0 | #5 |

8. Complete the following excitation table for JK flip-flop:

| Q(t) | Q(t+1) | J | K |
|---|---|---|---|
| 0 | 0 | 0 | d |
| 0 | 1 | 1 | d |
| 1 | 0 | d | 1 |
| 1 | 1 | d | 0 |

10. Find the state table for the following state diagram:

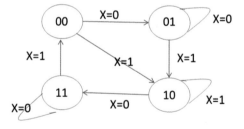

| Present state | | Next state | | |
|---|---|---|---|---|
| | | X = 0 | | X = 1 |
| A | B | A | B | AB |
| 0 | 0 | 0 | 1 | 10 |
| 0 | 1 | 0 | 1 | 10 |
| 1 | 0 | 1 | 1 | 10 |
| 1 | 1 | 1 | 1 | 00 |

12. Show state table and state diagram for the following circuit:

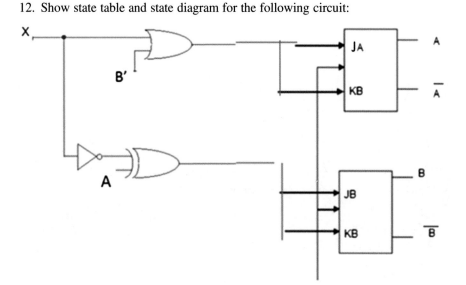

| Present state | | Next state | | | | Clock |
| --- | --- | --- | --- | --- | --- | --- |
| | | X = 0 | | X = 1 | | |
| A | B | A | B | A | B | |
| 0 | 0 | 1 | 1 | 1 | 0 | |
| 0 | 1 | 0 | 1 | 1 | 1 | |
| 1 | 0 | 0 | 0 | 0 | 1 | |
| 1 | 1 | 1 | 1 | 0 | 0 | |

# Chapter 6

## *Review Questions*

Multiple Choice Questions

The function of the _____ is to perform arithmetic operations.

(a) Bus
(b) Serial port
(c) ALU
(d) Control unit

Answer: C

2. When you compare the functions of a CPU and a microprocessor, _____

   (a) They are the same.
   (b) They are not the same.

(c) The CPU is faster than microprocessor.
(d) The microprocessor is faster than CPU.

Answer: A

4. The CISC processor control unit is _____.

(a) Hardware
(b) Microcode
(c) (a) and (b)
(d) None of the above

Answer: A

6. Which of the following buses are 32-bit?

(a) ISA
(b) PCI and EISA
(c) EISA and ISA
(d) MCA and ISA

Answer: A

8. How many memory location does have a memory with 16 Address Lines

(a) 1K
(b) 4K
(c) 64K
(d) 32K

Answer: AC

10. The Fetch instruction means

(a) Executing Instruction
(b) Read Instruction from memory
(c) Decode Instruction
(d) Store Data

Answer: b

## Short-Answer Questions

2. What is function of OS?
   Operating systems manage computer hardware resources such as input/output operations, managing memory, and scheduling processes for execution
4. What is function of Assembler?
   Assembler converts assembly language to machine code.
6. List the components of a microcomputer.

Answer: The components of a microcomputer consist of the following: microprocessor(CPU), buses, memory, serial input/output, programmable I/O interrupt, and direct memory access (DMA).

8. List the functions of an ALU.

Answer: The function of the ALU is to perform arithmetic operations such as addition, subtraction, division and multiplication, and logic operations such as AND, OR and NOT.

10. List components of a CPU?

Answer: The components of a CPU are the following: arithmetic logic unit (ALU), control unit, and registers.

12. How many bits is a word?

32

14. Explain the function of DMA.

Answer: DMA (direct memory access) allows the transfer of blocks of data from memory to an I/O device or vice versa. This is done directly without using the CPU.

Answer: If the control unit registers and ALU are packaged into one integrated circuit, it is a microprocessor; if they are not packaged in the same unit, it is a CPU.

16. What is the application of a serial port?

Answer: USB which has many applications is a type of serial port.

18. List the types of memory used in a computer.

Cache, Main Memory, and Disk

20. What is type of memory use for main memory?

DRAM

22. What are the characteristics of a 32-bit machine?

32- bit machine has 32-bit registers, 32-bit ALU and perform 32-bit operations

24. List characteristics of CISC processor

Answer: The characteristics of a CISC processor are the following: A large number of instructions, many addressing modes, variable length of instructions, most instructions can manipulate operands in the memory and control unit is microprogrammed.

26. Distinguish between von Neumann Architecture and Harvard architecture

Answer: In the von Neumann Architecture instructions are sent over the data bus while the Harvard Architecture uses separate buses for data and instructions.

28. List CPU instruction execution steps

Answer: The CPU instruction and execution steps are the following: Fetch, Decode, Execute, and Write.

30. Explain the decode instruction.

Answer: The control unit determines the type of instruction and move the contents of registers to the input of ALU

32. What does IR stand for and its application

Answer: Instruction Register and holds instruction

34. List types of disk controller

Answer: There are two types of disk controllers being integrated disk electronics (IDE) and small computer system interface (SCSI).

36. List the two most computer Buses

Answer: PCI-64 and PCI-express

40. Show diagram of PCIe lane

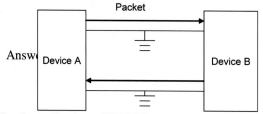

42. What is application of HDMI?

Answer: HDMI is used to transfer uncompressed video data and compressed or uncompressed digital audio signals from one device or another. This is used for computer monitors, TVs, and video projectors.

## Chapter 7: Questions and Problems

2. What does RAM stand for?

Answer: RAM stands for random-access memory.

4. Which of the following memory types are used for main memory?

(a) ROM and SDRAM
(b) SRAM and DRAM
(c) SDRAM and DRAM
(d) DRAM and EPROM

Answer: C

6. What does ROM stand for?

Answer: Read-only memory

8. What is the difference between EEPROM and EPROM?

Answer: EPROM requires the use of an ultraviolet light to be erased, while EEPROM can be erased by applying a specific voltage to one of its pins.

10. What is the primary application of SRAM?

Answer: The primary application of SRAM is used as cache for the CPU.

12. Define the following terms:

(a) Track
(b) Sector
(c) Cluster

Answer:

(a) Tracks are the division of platters into circular paths.

(b) Sectors are each of the tracks further broken down into smaller pieces.

(c) A cluster is a grouping of sectors.

14. What is the function of File Allocation Table (FAT)?

Answer: FAT defines organization of information stored on a hard disk, FAT16 and FAT32 were used on earlier Windows applications.

16. What are the types of cache?

Answer: Data cache (D-cache) and Instruction cache (I-cache).

18. What is virtual memory?

Answer: Virtual memory is either a hard disk drive (HDD) or solid-state drive (SSD) that is used to store application data and instructions from the main memory that are not currently needed by the CPU.

20. Physical address determines the size of

(a) Virtual memory

(b) Physical memory

(c) Cache memory

Answer: B

22. What is hit ratio?

Answer: Hit ratio is the number of hits/number of misses number of hits.

24. Explain spatial locality.

Answer: The idea that when a memory location is accessed, it is very likely that nearby memory locations will also need to be accessed.

26. Show a format of address seen by the cache for direct mapping.

Answer:

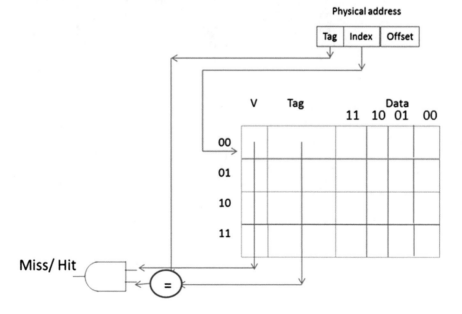

28. What is the function of a page number in a virtual address?
    Answer: The page number is used as part of a virtual address to identify pages.
30. What is the function of the page table?
    Answer: The page table is used to keep track of the page number of each page
    and the corresponding block of data. The page table also keeps track of whether
    each page is in the main memory or virtual memory.
32. What is the advantage of set associative versus direct mapping of caches?
    Answer: Set associative mapping will have less misses than direct mapping.
34. _____ is the fastest type of memory.

    (a) Cache memory
    (b) Main memory
    (c) Secondary memory
    (d) Hard disk

    Answer: A


Problems

2. The following memory and cache memory are given. CPU generates addresses
   0x1, 0x2, 0x1, 0x8, 0x9, 0x1C, 0x1D, 0x3, and 0x4.

   (a) Show the contents of the cache using two-way set associative mapping;
       assume a LRU replacement policy.
   (b) What is the hit rate?

| Address | Content | Address | Content |
|---------|---------|---------|---------|
| 00000   | 5       | 10000   | 5       |
| 00001   | 3       | 10001   | 0       |
| 00010   | 11      | 10010   | 1       |
| 00011   | 6       | 10011   | 11      |
| 00100   | 7       | 10100   | 15      |
| 00101   | 8       | 10101   | 09      |
| 00110   | 9       | 10110   | 12      |
| 00111   | 12      | 10111   | 23      |
| 01000   | 0       | 11000   | 65      |
| 01001   | 0       | 11001   | 21      |
| 01010   | 8       | 11010   | 8       |
| 01011   | 7       | 11011   | 7       |
| 01100   | 9       | 11100   | 9       |
| 01101   | 0       | 11101   | 0       |
| O1110   | 2       | 11110   | 2       |

01111   5          11111   5

Initial value for V-bit and LUR

| Set address | V | Tag | B1 | B0 | LRU | v | Tag | B1 | B0 | LRU |
|---|---|---|---|---|---|---|---|---|---|---|
| 00 | 0 | | | 0 | 0 | | | | | 0 |
| 01 | 0 | | | 0 | 0 | | | | | 0 |
| 10 | 0 | | | 0 | 0 | | | | | 0 |
| 11 | 0 | | | 0 | 0 | | | | | 0 |

Format of address seen by cache

| 2 bits | 2 bits | 1 bit |
|---|---|---|
| Tag | Set address | Byte offset |

| Set address | V | Tag | B1 | B0 | LRU | V | Tag | B1 | B0 | LRU |
|---|---|---|---|---|---|---|---|---|---|---|
| 00 | 0–>1 | 00 | 0 | 5 | 0–>–l>–0 | 0–>1 | 01 | 0 | 5 | 0–>1 |
| 01 | 0–>1 | 00 | 11 | 3 | 0–>1 | 0–>1 | | | | 0 |
| 10 | 0–>1 | 11 | 0 | 9 | 0–>–1>–0 | 0–>1 | 00 | 7 | 6 | 0–>1 |
| 11 | 0 | | | | 0 | 0 | | | | 0 |

4. A computer has 32 Kbytes of virtual memory and 8 Kbytes of main memory with a page size of 512 bytes.

   (a) How many bits are in the virtual address?
   (b) How many pages are in the virtual memory?
   (c) How many bits are required for the physical address?
   (d) How many frames or blocks are in the main memory?

   (a) 15 virtual address
   (b) $2^{15} / 2^9 = 2^6 = 64$ pages
   (c) $2^{13} = 8k$ physical address is 13 bits

6. A computer has 20 bits of virtual memory and each page is 2KB.

   (a) What is the size of virtual memory?
   (b) How many pages are in virtual memory?

**Answer:**

(a) $2^{20} = 1MB$
(b) $2^{20} / 2^{11} = 2^9$ pages

8. CPU of Fig. 7.24 generates addresses 0x00 and 0x0b; assumes Page0 map into block1 and page 2 map in block 0; shows the contents of page table

   Address 0X00

| Page# | Offset |
|---|---|
| 000 | 00 |

Page number is the address to the page table
Address 0X0b

|  | Page table | |
|---|---|---|
|  | Valid bit | Frame number (2 bits) |
| 000 | 0 1 | 01 |
| 001 | 0 | |
| 010 | 0 1 | 00 |
| 011 | 0 | |
| 100 | 0 | |
| 101 | 0 | |
| 110 | 0 | |
| 111 | 0 | |
|  | 0 | |
| Page # | offset | |
| 010 | 11 | |

010 is the address to page table.

# Chapter 8

2. List types of instructions based on number of operands:

   Instructions with no operand such as HLT
   Instruction with two operands such as MOV R1, R2
   Instruction with the operands such as ADD R1, R2, R3

4. Which register of ARM processor is used for Stack Pointer (SP)?

   R13

6. What is contents of R5 after execution of the following instruction, assume R2 contains 0X34560701 and R3 contains 0X56745670

   1. ADD R5, R2, R3
   2. AND R5, R3, R2
   3. EOR R5, R2, R3
   4. ADD R5, R3, #0x45

      (a) 0x8ACA5D71
      (b) 0x04540600
      (c) C.0x32225171
      (d) 0x567456B5

8. What is contents of R5 after execution of the following instruction, assume R2 contains 0X34560701 and R3 contains 0X56745670

(a) ADD R5, R2, R3
    R5=0x8ACA5D71
(b) AND R5, R3, R2
     R5=0x14540600
(c) EOR R5, R2, R3
      R5=0x66225171
(d) ADD R5, R3, #0x45
      R5=0X567456B5

10. What is contents of R3

MOV R1, #0x52
LSL R3, R1, #0x8 R3=0x5200

12. What is the difference between these two instructions?

(a) SUBS R1, R2, R2
(b) SUB R1, R2, R2

        Question a does not change bits in PSR Register,
        Question b will change bits in PSR

14. Trace the following instructions

MOV R1, #0x0F
MOV R2, #0x23
AND R4, R2, R1 R4= 0x03

16. What are the contents of R1? Assume R2 = 0x00001234.

(a) MOV R1, R2, LSL #4
     R1 = 0x00012340
(b) MOV R1, R2, LSR #4
     R1 = 0x00000123

18. What is contents of R1 after executing the following Instruction assume R1=0xF1245678

ROR R1, R1, #8 R1=0x78F12456

20. Convert the following HLL language to ARM instructions.

```
    IF R1>R2 OR R3>R4 then
    R1= R1 +1
    Else
    R3=R3 +R5*8
    Endif
  CMP R1, R2
  CMPLE R3, R4
```

```
ADDGT R1, R1,#01
ADDLE R3, R3, R4, LSL, #3
```

22. Write a program to add ten numbers from 0 to 10 or convert the following C language to ARM assembly language

```
    int  sum;
  int i;
  sum = 0;
  for (i = 10 ; i > 0 ; i - - ){
  sum = sum +1
  }
```

**Solution**
R5 Hold the Sum
R10 holds the i
SUB R5, R2, R2 ;clear the sum
MOV R10, #10
Loop
CMP R10, 0
ADDGT R5, R5,#1
SUBGT R10,R10, #1
BGT Loop

24. Convert the following flow chart to ARM assembly:

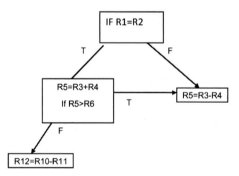

```
    CMP R1, R2
    SUBNE R5, R3,R4
    BNE Halt
    ADDEQ R5, R3 R4
    CMPEQ R5, R6
 SUBLE R12, R10, R11
 SUBGT R5, R3,R4
 Halt
```

# Chapter 9

Problem

2. Show how the following list is stored in memory using Little Endian
   List DCW 0x534, 0x22, 0x167,0x5692

| | |
|---|---|
| List | 34 |
| List +1 | 05 |
| List +2 | 22 |
| List +3 | 00 |
| List+4 | 67 |
| List+5 | 01 |
| List +6 | 92 |
| List+7 | 56 |

4. What is the application of the ADR instruction?
   ADR use to transfer address of a label into a register
   ARR R0, List
6. Why must a character string be terminated by a null character in ARM assembly?
   To indicates the end of string
8. Write assembly language for the following HLL

```
IF R1 = R0
    Then
ADD R3, R0, #5
Else
SUB R3, R0, #5
Solution
      MOV R0, # N1
      MOV R1, # M1
__main
      CMP R1,R0
      ADDEQ R3,R0,#0x5
      SUBNE R3, R1,#0x05
```

10. Write a program to multiplying two number assume multiplicand is 0x2222222
    and multiplier is 3, check your result with µVision

```
__main
      LDR R1, =0x22222222
      LDR R2, =0x3
      MUL R3, R1, R2
```

12. Write a subroutine to calculate value of Y where $Y = X^{*2} + x + 5$, assume X represented by

```
N1  EQU  0x5
Solution
N1  EQU   0x5
__main
    LDR R1, =N1
    BL  Fun
    MOV R7, R2
Fun
    MUL R2, R1, R1
    ADD R2, R2, R1
    ADD R2 , R2, # 0x5
    BX R14   ; Return to main program
```

14. Write a program to read a word memory location LIST and Clear bit position B4 through B7 of register R5 , assume R5 contains 0XFFFFFFFF

```
LDR R0, =0x000000F0
LDR R5, =0xFFFFFFFF
Solution
    LDR R0, =0x000000F0
      LDR R5, =0xFFFFFFFF
__main
    BIC R4, R5, R0
```

# Chapter 10

**Solutions**
 For all program, the following templet were used:
```
    AREA RESET, DATA, READONLY
    EXPORT __main
        ENTRY
__maim
        Program Code
STOP B STOP
    END ;End of the program
```
 Default memory location 0x20000000 through 0x20020000 is reserved for writing and reading
 For storing data ay memory location list  the address of list1 must be added to 0x20000000.
```
ADR R0, List1
MOV R2, #0x20000000
ADD R0, R2, R0
R0 will hold the address of List1
```

2. Work problem #1 part a and b using Little Endian.

   (a) R1 = 0xE532F534
   (b) R2 = 0xFE080201

4. What are the contents of register Ri for the following load Instructions? Assume R0 holds the address of list using Little Endian.

   (a) LDR R1, [R0]
   (b) LDRH R2, [R0]
   (c) LDRB R3, [R0], #1
   (d) LDRB R4, [R0]
   (e) LDRSB R5, [R0], #1
   (f) LDRSH R6, [R0]

   List DCB   0x34, 0xF5, 0x32, 0xE5, 0x01, 0x02
   **Solution**

   (a) LDR R1, [R0]; R1=0xE532F534
   (b) LDRH R2, [R0]; R2=0x0000F534
   (c) LDRB R3, [R0], #1; R3=0x00000034
   (d) LDRB R4, [R0]; R4= 0x000000F5
   (e) LDRSB R5, [R0], #1; R5=0xFFFFFFF5
   (f) LDRSH R6, [R0]; R6=0xFFFFE532

8. Write a program to find the largest number and store it in memory location LIST3, Assume Numbers are in location LIST1 and LIST2

   List1 DCB 0x23, 0x45, 0X23, 0x11
   List2 DCB 0x0

```
__main
    ADR  R0,LIST1
    LDR R1, [R0]
    ADR R0, LIST2
    LDR  R2, [R0]
    CMP R1, R2
    BHI   RESULT        ; IF R1>R2
    MOV R1, R2
RESULT
    ADR R0, LIST3
      LDR R4,=0x20000000
       ADD R0, R4,R0
     STR R1, [R0]
LIST1   DCD   0x23456754
LIST2   DCD   0X34555555
LIST3   DCD   0x0
```

10. Write a program add LIST1 to LIST2 and store the sum in LIST3

```
__main
      ADRL R1, LIST1     ; ADR is Pseudo Instruction
    LDR R2, [R1]
    ADR R3, LIST2
    LDR R4, [R3]
    ADD R5, R2,R4
    ADR R8, LIST3
    STR R5 , [R8]
LIST1   DCD 0x00002345
LIST2   DCD 0X00011111
LIST3   DCD 0x00000000
```

12. Write a program to add 8 numbers using Post-Index addressing
       LIST DCB 0x5, 0x2, 0x6, 0x7, 0x9, 0x1, 0x2, 0x08

```
Solution:
     ADR R0, LIST
__main
    SUB R5,R5,R5
    MOV R1,#0x8
LOOP
    LDRB R2, [R0] , #1
    ADD R5, R5, R2
    SUB R1, R1, #01
    CMP R1, #0x0
    BNE   LOOP
    ALIGN
LIST DCB 0x5, 0x2,0x6,0x7 ,0x9,0x1,0x2,0x08
```

14. Write a program to convert the following HLL to assembly language

```
If  R1=R2 then
R3= R3+1
IF R1<R2 Then
R3=R3-1
If R1>R2 Then
R3=R3-5
   Solution

   MOV R1, #0x9
   MOV R2, #0x6
       MOV R3, #0x5
    __main
     CMP R1, R2
     ADDEQ R3, R3, #0x1
     SUBLE R3, R3, #0x1
ADDGT R3, R3, #0x3
```

16. Write a program to compare two numbers and store largest number in a memory location LIST

```
__main
      MOV R1 , #M1
      MOV R2, #N1
      CMP R1, R2
      MOVGT R3, R1
      MOVLE R3, R2
      ADR R0, LIST2
      LDR R5,=0x20000000
      ADD R0, R0, R5
   STRB R3,[R0]
M1     EQU  5
N1     EQU  6
LIST2  DCB 0x0
```

18. Convert the following ARM instruction to machine code

(a) ADD R5, R6, R8

| 31      28 | 27 26 | 25 | 24      21 | 20 | 19      16 | 15      12 | 11      0 |
|---|---|---|---|---|---|---|---|
| Cond | 0 0 | I | Op code | S | Rn | RD | Operand 2 |

| 1110 00 | 0 | 0100 | 0 | 0110 | 0101 | 1000 |
|---|---|---|---|---|---|---|

(b) ADDNE R2, R3, 0x25

| 31      28 | 27 26 | 25 | 24      21 | 20 | 19      16 | 15      12 | 11      0 |
|---|---|---|---|---|---|---|---|
| Cond | 0 0 | I | Op code | S | Rn | RD | Operand 2 |

(c) BNE label

| 31      28 | 27      25 | 24 | 0 |
|---|---|---|---|
| Cond | 101 | L | offset |
| 0001 | 101 | 0 | offset |

# Chapter 11

2. Write a program to set bit position 15 in C and assembly

```
C Language
int main(void){
int x = 0;
  x |= (1 << 15);
  return (0);
}
Assembly
EXPORT __main
  ENTRY
__main
     ; set b15 to one
    MOV R1, #0X00
    ORR R1, #0x8000
STOP   B   STOP
          END
```

4. Write a program to clear position 15 in C and assembly

```
int main(void){
  int x = 0xffffff;
  x &= ~ (1 << 15));
  return (0);
}
      AREA   MYCODE, READONLY, CODE
   EXPORT __main
     ENTRY
__main
   LDR R1,=0xFFFFFFFF
   AND R1, #0xFFFF7FFF; clear bit b15
stop   b stop
       END
```

6. Write a program to shift left contents register 8 times in C and assembly, assume register holds 0x4

```
int main(void){
  int x = 0x4;
    x = x<<8;
  return (0);
{
```

```
AREA   MYCODE, READONLY, CODE
  EXPORT __main
    ENTRY
__main
   MOV R1, #0x4
```

```
  LSL R1, R1, # 0x08
stop    b stop
        END
```

8. Write a program two swap high digit with low digit of 0x45 using C and assembly

**C Programming**
```
  int main (void) {
    /* local variable definition */
      int X = 0x45;
      int Y;
      Y = X & 0x0F;
      Y = Y<< 4;
      X = X >> 4;
      X = X | Y;
  return 0;
}
```
**Assembly**
```
EXPORT __main
        ENTRY
__main
  MOV R1, #0x45
  MOV R2, R1
  AND R2, #0xF0
  LSR R2, #4
  AND R1, #0x0F
  LSL R1 , #0x4
  ORR R1, R2,R1
done      B done
        END
```

## References

1. M. Mano, *Digital Design*, 5th edn. (Pearson, 2013)
2. E.O. Hwang, *Digital Design and Microprocessor Design with Interfacing*, 2nd edn. (Cengage Learning, 2018)
3. D. Haris, S. Haris, *Digital Design and Computer Architecture. ARM Edition* (Morgan Kaufmann, 2016)
4. M. Wolf, *Computers as Components* (Morgan Kaufmann, 2017)
5. A. Elahi, T. Arjeski, *ARM Assembly Language with Hardware Experiments* (Springer, 2015)
6. W. Stalling, *Computer Organization and Design*, 10th edn. (Pearson, 2016) USA
7. A. Clements, *Computer Organization and Architecture Themes and Variations* (Cengage Learning, 2014)

8. NXP Corp., LPC16XX User Manual
9. http://infocenter.arm.com, ARM V7 Manual
10. Keil Corp., μvision Development Tool
11. ARM Cortex-M3 Technical Reference Manual
12. S.B. Furber, *ARM System-on-Chip Architecture* (Addison Wesley, 2000)
13. W. Holm, *ARM Assembly Language* (CRC Press, 2009)
14. K. Schindler, *Introduction to Microprocessor Based System Using the ARM Processor* (Person, 2013)
15. J.W. Valvano, *Embedded Systems Real-time Interfacing to the ARM Cortex-M3* (J.W. Valvano, 2011)
16. D. Lewis, *Fundamentals of Embedded Software with ARM Cotex-M3* (Pearson, 2013)
17. R. Gibson, *ARM Assembly Language—An Introduction* (LuLu, 2007)

# Bibliography

## Chapter 1

1. A. Elahi, *Computer Systems: Digital Design Fundamental of Computer Architecture and Assembly Language* (Springer, Cham, 2018)
2. A. Elahi, T. Arjeski, *ARM Assembly Language with Hardware Experiments* (Springer, New York, 2015)
3. D. Patterson, J. Hennessy, *Computer Organization and Design, The Hardware/Software Interface* (Morgan Kaufmann, Burlington, 2011)
4. L. Null, J. Lobur, *The Essentials of Computer Organization and Architecture* (Jones & Bartlett Learning, Burlington, 2014)
5. C. Hamacher, Z. Vranesic, S. Zaky, *Computer Organization*, 5th edn. (McGraw-Hill, New York, 2002)
6. U. Ramachandran, W.D. Leahy, *Computer Systems an Integrated Approach to Architecture and Operating Systems* (Pearson, London, 2011)
7. M. Mano, *Digital Design*, 5th edn. (Pearson, 2013)
8. E.O. Hwang, *Digital Design and Microprocessor Design with Interfacing*, 2nd edn. (Cengage Learning, Boston, 2018)
9. D. Haris, S. Harris, *Digital Design and Computer Architecture ARM Edition* (Morgan Kaufmann, Burlington, 2016)

## Chapter 2

10. 1. M. Mano, *Digital Design*, 5th edn. (Pearson, 2013)
2. A. Elahi, T. Arjeski, *ARM Assembly Language with Hardware Experiments* (Springer, New York, 2015)
3. A. Elahi, *Computer Systems: Digital Design Fundamental of Computer architecture and Assembly Language* (Springer, Cham, 2018)
4. D. Haris, S. Haris, *Digital Design and Computer Architecture ARM Edition* (Morgan Kaufmann, Burlington, 2016)
5. C. Unsalan, B. Tar, *Digital System Design with FPGA: Implementation Using Verilog and VHDL* (McGraw-Hill Education, New York, 2017)
6. P. Halmos, S. Givant, *Introduction to Boolean Algebras* (Springer, New York, 2009)

© The Editor(s) (if applicable) and The Author(s), under exclusive license to
Springer Nature Switzerland AG 2022
A. Elahi, *Computer Systems*, https://doi.org/10.1007/978-3-030-93449-1

# Chapter 3

16. A. Elahi, *Computer Systems: Digital Design Fundamental of Computer Architecture and Assembly Language* (Springer, Cham, 2018)
17. A. Elahi, T. Arjeski, *ARM Assembly Language with Hardware Experiments* (Springer, New York, 2015)
3. M. Mano, *Digital Design*, 5th edn. (Pearson, 2013)
4. M. Rafiguzzaman, *Fundamentals of Digital Logic and Microcontroller* (Wiley, Hoboken, 2005)
5. E.O. Hwang, *Digital Design and Microprocessor Design with Interfacing*, 2nd edn. (Cengage Learning, Boston, 2018)
6. D. Haris, S. Haris, *Digital Design and Computer Architecture ARM Edition* (Morgan Kaufmann, Burlington, 2016)
7. U. Ramachadrean, W. Leahy, *Computer Systems an Integrated Approach to Architecture and Operating System* (Addison Wesley, Boston, 2011)
8. C. Unsalan, B. Tar, *Digital System Design with FPGA: Implementation Using Verilog and VHDL* (McGraw-Hill Education, New York, 2017)


# Chapter 4

24. A. Elahi, *Computer Systems: Digital Design Fundamental of Computer Architecture and Assembly Language* (Springer, Cham, 2018)
2. M. Mano, *Digital Design*, 5th edn. (Pearson, 2013)
3. Logisim. https://www.cburch.com/logisim/.
4. M. Rafiguzzaman, *Fundamentals of Digital Logic and Microcontroller* (Wiley, Hoboken, 2005)
5. E.O. Hwang, *Digital design and microprocessor design with interfacing*, 2nd edn. (Cengage Learning, Boston, 2018)
6. D. Haris, S. Haris, *Digital Design and Computer Architecture. ARM Edition* (Morgan Kaufmann, Burlington, 2016)
7. U. Ramachadrean, W. Leahy, *Computer Systems an Integrated Approach to Architecture and Operating System* (Addison Wesley, Boston, 2011)


# Chapter 5

31. 1. M. Mano, *Digital Design*, 5th edn. (Pearson, 2013)
2. Logisim. https://www.cburch.com/logisim/.
3. M. Rafiguzzaman, *Fundamentals of Digital Logic and Microcontroller* (Wiley, Hoboken, 2005)
4. E.O. Hwang, *Digital Design and Microprocessor Design with Interfacing*, 2nd edn. (Cengage Learning, Boston, 2018)
5. S. Haris, D. Haris, *Digital Design and Computer Architecture: ARM Edition* (Morgan Kaufmann, Burlington, 2016)
6. U. Ramachadrean, W. Leahy, *Computer Systems an Integrated Approach to Architecture and Operating System* (Addison Wesley, Boston, 2011)

# Chapter 6

1. A. Elahi, *Computer Systems: Digital Design Fundamental of Computer Architecture and Assembly Language* (Springer, Cham, 2018)
101. A. Elahi, T. Arjeski, *ARM Assembly Language with Hardware Experiments* (Springer, New York, 2015)
3. W. Stallings, *Computer Organization and Architecture*, 9th edn. (Pearson, 2012)
4. U. Ramachandran, W.D. Leahy, *Computer Systems an Integrated Approach to Architecture and Operating Systems* (Pearson, London, 2011)
5. D. Patterson, J. Hennessy, *Computer Organization and Design, The Hardware/Software Interface* (Morgan Kaufmann, Burlington, 2011)
6. L. Null, J. Lobur, *The Essentials of Computer Organization and Architecture* (Jones & Bartlett Learning, Burlington, 2014)
7. C. Hamacher, Z. Vranesic, S. Zaky, *Computer Organization*, 5th edn. (McGraw-Hill, New York, 2002)

# Chapter 7

44. A. Elahi, *Computer Systems: Digital Design Fundamental of Computer Architecture and Assembly Language* (Springer, Cham, 2018)
011. A. Elahi, T. Arjeski, *ARM Assembly Language with Hardware Experiments* (Springer, New York, 2015)
3. D. Anderson, T. Shanley, *Pentium Pro and Pentium II Architecture* (Addison-Wesley, Boston, 1998)
4. W. Stallings, *Computer Organization and Architecture*, 9th edn. (Pearson, 2012)
5. U. Ramachandran, W.D. Leahy, *Computer Systems an Integrated Approach to Architecture and Operating Systems* (Pearson, London, 2011)
6. D. Patterson, J. Hennessy, *Computer Organization and Design, The Hardware/Software Interface* (Morgan Kaufmann, Burlington, 2011)
7. L. Null, J. Lobur, *The Essentials of Computer Organization and Architecture* (Jones & Bartlett Learning, Burlington, 2014)

# Chapter 8

1. A. Elahi, T. Arjeski, *ARM Assembly Language with Hardware Experiments* (Springer, New York, 2015)
2. S. Fuber, *ARM System-On-Chip Architecture* (Addison-Wesly, Boston, 2000)
3. J.W. Valvano, *Introduction to ARM Cortex-M Microcontrollers Embedded Systems* (Jonathan W. Valvano, 2013)
4. J. Yiu, *The Definitive Guide to the ARM Cotex-M3* (Newnes, London, 2010)
5. Y. Bai, *Pracrical Microcontroller Engineering with ARM Technology* (IEEE Press, Piscataway, 2016)
6. R. Gibson, *ARM Assembly Language—An Introduction.* (LuLu, 2007)
57. ARM Limited. https://www.arm.com
8. ARM Limited. ARMv7 Architecture Reference Manual. (2008)

# Chapter 9

59. A. Elahi, *Computer Systems: Digital Design Fundamental of Computer Architecture and Assembly Language* (Springer, Cham, 2018)
60. A. Elahi, T. Arjeski, *ARM Assembly Language with Hardware Experiments* (Springer, New York, 2015)
3. MDK- ARM version 5.35. https://Keil.com
4. ARMv7-M Architecture Reference Manual. https://developer.arm.com
5. D. Lewis, *Fundamentals of Embedded Software with ARMCotex-M3* (Pearson, 2013)

# Chapter 10

1. A. Elahi, T. Arjeski, *ARM Assembly Language with Hardware Experiments* (Springer, New York, 2015)
2. L. Pyeatt, *Modern Assembly Language with the ARM Processor* (Elsevier, 2020)
3. S. Fuber, *ARM System-On-Chip Architecture* (Addiso-Wesly, Boston, 2000)
4. J.W. Valvano, *Introduction to ARM Cortex-M Microcontrollers Embedded Systems* (Jonathan W. Valvano, 2013)
5. Y. Bai, *Practical Microcontroller Engineering with ARM Technology* (IEEE Press, Piscataway, 2016)
6. D. Lewis, *Fundamentals of Embedded Software with ARMCotex-M3* (Pearson, 2013)
7. Y. Zhu, *Embedded Systems with ARM Cortex-M Microcontroller in Assembly language and C* (E-Man Press LLC, 2017)
9. ARM Limited. https://www.arm.com
10. ARM Limited. ARMv7 Architecture Reference Manual (2008)
11. J. Bakos, *Embedded Systems: ARM Programming and Optimization* (Morgan Kaufmann, Burlington, 2015)
12. Keil Embedded Development Tools for ARM, Cortex-M. https://www.keil.com

# Chapter 11

75. A. Elahi, T. Arjeski, *ARM Assembly Language with Hardware Experiments* (Springer, New York, 2015)
2. L. Pyeatt, *Modern Assembly Language Programming with ARM Processor* (Elsevier, 2020)
3. M. Mazidi, *Freescale ARM Cortex-M Embedded Programming: Using C Language*
4. D. Lewis, *Fundamentals of Embedded Software with ARMCotex-M3* (Pearson, 2013)
5. R. Gibson, *ARM Assembly Language—An Introduction.* (Lulu, 2007)
6. Y. Zhu, *Embedded Systems with ARM Cortex-M Microcontroller in Assembly language and C,* (E-Man Press LLC, 2017)
7. ARM Limited. https://www.arm.com
8. ARM Limited. ARMv7 Architecture Reference Manual (2008)
9. J. Bakos, *Embedded Systems: ARM Programming and Optimization* (Morgan Kaufman, Burlington, 2015)

# Index

Printed in the United States
by Baker & Taylor Publisher Services